CONTRIBUTIONS TO THE HISTORY OF STATISTICS

CONTRIBUTIONS TO THE HISTORY OF STATISTICS

BY

HARALD WESTERGAARD

FORMERLY PROFESSOR OF STATISTICS IN THE
UNIVERSITY OF COPENHAGEN

REPRINTS OF ECONOMIC CLASSICS

AUGUSTUS M. KELLEY · PUBLISHERS
NEW YORK 1969

First Edition 1932
(London: P. S. King & Son Ltd.,
Orchard House, Great Smith Street,
Westminster, S. W. 1, 1932)

Reprinted 1969 by
Augustus M. Kelley · Publishers
New York New York 10010

SBN 678 00521 4

Library of Congress Catalogue Card Number
70–85136

PRINTED IN THE UNITED STATES OF AMERICA
by SENTRY PRESS, NEW YORK, N. Y. 10019

PREFACE

In the following pages I have tried to sketch the evolution of statistics from its beginning up to the end of the past century. It is my hope that students of the history of statistics will find useful information in the volume, the material for studies of this kind being not always easily accessible, scattered as it is in the vast literature, in numerous monographs, in statistical journals and in official reports. By the choice of titles of reference I have wished to facilitate further study of the subjects concerned. But I am well aware that I shall not meet with general agreement as to the plan I have followed. Many readers will miss a detailed history of the "Staaten-kunde," to which I have only devoted a few pages. Among other works I may here refer to V. John, *Geschichte der Statistik* (1884), and to A. Gabaglio, *Teoria generale della statistica. Parte storica* (1888). Other readers may object that I ought to have dealt more fully with the Theory of Probabilities whereas I have only entered upon it where I found it had influence on the evolution of statistics. Others, again, will find official statistics treated in stepmotherly fashion; they may first of all be referred to the great co-operative work, *The History of Statistics*, published in 1918 by The American Statistical Association. Lastly, critics may regret that the rich evolution in the last decades is left out of consideration. It is, however, doubtful whether it is possible already to treat this epoch entirely objectively. It is even a question whether I have always succeeded in giving the history of the past centuries a strictly impersonal character, although I have tried to be as objective as possible. But the last thirty years form a most attractive period in the history of statistics, and if my advanced age permits it, I hope later to enter on the subject, attempting to value the present tendencies and to draw the horoscope of statistics on the basis of evolution in the present century.

I owe a word of thanks to many colleagues around in the world for kind assistance during my work, and also to several libraries to which I have had access; among these I may specially mention the library of the Life Insurance Company, Utrecht, in Holland.

Writing as I do in a foreign language, I am indebted to Miss H. Vernon Jones who kindly undertook the revision of the language. And, lastly, I have to thank the publishers, P. S. King & Son, for liberally accepting a book which can hardly expect much attention.

COPENHAGEN,
March, 1932.

CONTENTS

CHAPTER I

INTRODUCTION

1. I⊤ is always interesting to follow a branch of science from its first trifling beginning up to its present standpoint. It is like following a river from its apparently insignificant sources till it gradually becomes a mighty stream. A study of this kind will probably always teach us the same lesson. It will show us the human genius, struggling to find the truth under more or less favourable circumstances, frequently in conflict with apparently unsurmountable difficulties, often led astray, but never giving up until at last the truth—perhaps far more simple than was expected—has been found. And having reached this goal the human mind will follow a new track, seeking new difficulties and trying to throw light on new problems.

This exploration will generally show us numerous mistakes which nowadays may appear curious. As a matter of fact, these mistakes have often been the natural means towards finding the right solution, so that these apparent failures have, in reality, a due share in the progress.

It is tempting for a student of the evolution of any scientific system to confine himself to the few richly endowed men, with names remembered through generations or even centuries, who succeeded in mastering the chief problems. But this would be unjust. In the track of a Newton or a Pasteur followed numbers of less well equipped workers who have a just claim to our gratitude. They tilled the field which their ingenious teachers opened to them, thus doing work which it was necessary to perform as an inevitable condition if these discoveries were to be utilised to the benefit of mankind. It may be that we do not even know their names. This is probably very often the case in the history of statistics. We notice progress in the work of a statistical office, but it is hardly known to whom it should be ascribed, numbers of faithful workers having done their best to promote statistics, thus rendering mankind an important service even though their names will not be remembered outside of the narrow office rooms where their work was done.

2. Trying to find his way back to its origin, the student of the history of statistics may for a moment be puzzled by the curious change of names which has taken place. Taking old books down from the shelves of the libraries, he will find numbers of volumes which on their title-pages are called statistical, but which, judging by their contents, have little or nothing in common with statistics in the modern sense of the word. In fact, the name of statistics was in former days applied *to the comparative description of states* (what the Germans called "Staatenkunde"). This system can be traced as far back as Aristotle and was later cultivated by Italian and other authors until at last it reached its culmination in the German universities of the seventeenth and eighteenth centuries. Etymologists may find the root of the word "statistics" in the Italian word *stato*, and a *statista* would thus be a man who had to do with the affairs of the State. "Statistics" would consequently mean a collection of facts which might be of interest to a statesman, whether they were given in the form of numerical observations or not. As we shall see, a statistical description of the State in the modern sense of the word would have been an impossibility in those days, the required facts being either non-existent, or, at best, buried in Government archives. Gradually, however, this system approached real statistical descriptions, so that it will not be superfluous to give a brief sketch of its evolution.

A modern statistician will, however, feel much more

at home in the so-called *Political Arithmetic* which originated in England in the seventeenth century from observations on the lists of births and deaths. First of all, it took up questions of mortality and other problems of vital statistics, later also turning to economic statistics. Gradually the expression "Political Arithmetic" grew obsolete, and the name statistics, which the German professors had used for their description of States, was adopted in its place.

But besides these two lines of evolution, the *Calculus of Probabilities* deserves attention. From an apparently insignificant discussion of the results of games of dice, etc., this system reached a very high development. It will suffice to mention as a standard work Laplace's masterly *Théorie analytique des probabilités* (1812). For a long while, however, the calculus of probabilities had less influence on statistics than might have been expected, the authors chiefly confining themselves to abstract theories which had little or nothing to do with reality.

I shall try in the following pages to give a bird's-eye view of these three lines of evolution, beginning with a brief sketch of the history of the comparative description of States.

CHAPTER II

THE DESCRIPTION OF STATES

3. In tracing the history of the "Staatenkunde" back to its origin, we first meet the name of Aristotle. His *Politeiai* contained monographs of no less than 158 States. Most of these descriptions are lost, but the existing monograph of Athens will give an idea of his plan and method. He gave the history of each of these mostly diminutive States with a description of its present character, referring also to its relations to the neighbouring States. There were details concerning public administration, justice, science and arts, religious life, manners and customs, etc. This collection of monographs was intended as a contribution to his famous theory of the State. From this point of view alone they might claim rank in the literature of the time.

Quite naturally the awakening political life in various Italian cities in the beginning of the modern era gave rise to similar compilations. Two well-known authors may be quoted here, viz. Francesco Sansovino (1521–86) and Giovanni Botero (1540–1617).

Sansovino published a work, *Del governo e amministrazione di diversi regni e republiche* (Venice, 1562, republished several times). He describes 22 States, among which were ancient Sparta, Athens and Rome. He even devoted a chapter to "Utopia." We are, of course, not entitled to expect an exhaustive description, even though Sansovino seems to have made earnest efforts to collect as many facts as possible. In the chapter about England one page only is devoted to the income of the State, eight lines to Justice, etc. On the other hand, the reader will find particulars regarding the

Order of the Garter, Parliament, the Coronation of the King and the order of Succession.

Giovanni Botero published in 1593 *Le relazioni universali*, a work with similar contents, though differently arranged. He deals with his subject from various points of view. He devotes part of his work to a description, mainly geographical, of the territories of the States. In a second part he gives an outline of their constitution and of the causes of their magnitude and wealth, and lastly, deals with the religious conditions of each individual country.

In France, also, there were interesting authors who may justly be included in this group. One of these is Etienne Pasquier (1529–1615), an advocate at the Parliament of Paris. He published in 1581 his famous *Recherches de la France*, a most interesting work, though his critics have objected that it lacks both plan and method. He deals with the legal problems, particularly from a national point of view. He tries to understand the history of France, to trace the cause of origin of the French Empire, and to give a clear conception of the structure of Society, of the Courts and other Institutions.

Of a little later date is a work published by Pierre d'Avity (1573–1635), *Les Estats, Empires, Royaumes, Seigneuries, Duchez et Principautez du Monde* (1614). On the whole it reminds one of its Italian predecessors, although perhaps the contents are of a more geographical nature. He describes in some detail manners and customs in previous times as well as among his contemporaries; he deals with "Genealogy" of the successive Princes in the countries which he describes, giving lists of their names; he gives details with regard to Succession and Coronation, the Court, the Nobility (with lists of the members of the highest rank). Further, he writes about "richesses," viz. currency and income of the State, also about religion and military matters. Quite naturally there are very few numerical statements.

A little later a series of thirty-six small volumes appeared in Holland (1624–40), published by the brothers Abraham and Bonaventura Elzevir. The books became known as the "Small republics of the Elzevires." They were written by J. de Laet (1590–1649), in co-operation with several other authors, and gave a description of the various countries, their climate and soil, the manners and customs of the population, the military forces, Succession of Princes, etc.

4. Thus a whole literature had been founded, more or less containing material for a comparison of various States.

It was in Germany that this system found its principal home. The works in question are, however, of a later date than those quoted above. Earlier works which might be expected to belong to this group have generally another point of view, as, for instance, the so-called cosmographies by Sebastian Franck and Sebastian Münster (1534 and 1540), which were principally of a geographical character, based on the very incomplete observations which were available, and on the whole very naïve. A student of the comparative description of States will also in vain turn over the leaves of the volumes of the *Theatrum Europæum*, begun by J. P. Abelin, and continued through a century with chronicles for each year from 1617. The first volume was published in 1635 for the years 1617–29.

Of more interest in this respect is the famous *Teutscher Fürstenstaat* (1656), which Seckendorff published (1626–92). This book, afterwards quoted an infinite number of times, gave an outline of the constitution and public administration of a normal German state, being especially written for the benefit of young noblemen who wished to qualify themselves for public service in one of the numerous small States of which Germany consisted. The German authors who in the following period took up the comparative description of States seem frequently to have used this book as a model.

The long series of German University Professors who cultivated the system of "Staatenkunde" was opened by the well-known polyhistor Hermann Conring (1606–

81), who lived in Helmstedt in Brunswick, where in that period there was a flourishing university. From 1660 he lectured regularly to a large audience at his home on this system. He had no manuscript for his lectures, relying only on his unique memory. Some of his disciples, however, made notes, and one of these (Oldenburger) published an abstract of Conring's lectures (1675), although much to his discontent. Conring's works on these subjects were published after his death (1730), from notes made by other disciples.[1]

Conring describes Spain with its Colonies, Portugal, France, etc., and finally Japan, Morocco and Abyssinia. The description of Japan covers only a few pages, the main sources of information being reports from the Jesuits. On the whole, there is in Conring's lectures much which recalls his predecessors. He has numerous quotations but no numerical observations, confining himself to vague expressions (that a country is populous, etc.). Like the scholastic philosophers, he arranges his subjects according to certain principles, choosing the following four: *causa materialis, finalis, formalis* and *efficiens*. Under the *materia* of the States he treats Territory and Population, under *forma*, Constitution and Administration, under *causa finalis* the aims of the State, and under *causa efficiens* Finance, and the Army and Navy.[2]

5. A number of German professors, chiefly jurists, followed the trail. One of these, Martin Schmeitzel, who died 1747, called his lectures *Collegium politico-statisticum*, and from this time the name of statistics came into use, especially through Gottfried Achenwall (1719–72), who died as professor at the University of Göttingen.

---

[1] This edition of Conring's works on Law and Political Science was published by Goebel in six folio volumes. The fourth volume contains the lectures on "Staatenkunde": Hermanni Conringii Examen Rerumpublicarum potiorum totius orbis. Opus ex tribus codicibus manuscriptis longe exactius editione Oldenburgeriana adornatum in quo respublicæ sequenti ordine illustrantur. . . .

[2] Compare John, *Geschichte der Statistik*, Stuttgart, 1884, pp. *58 sq.*; Zehrfeld, *Hermann Conrings Staatenkunde*, Berlin and Leipzig, 1926.

Achenwall defines statistics as the description of the Constitution in a wider sense, viz. the whole complex of remarkable facts regarding a State ("Staatsmerkwürdig-keiten ").[1]   But for practical reasons he will confine himself to the strictly necessary facts, choosing only the most important ones, which can throw light on the whole organisation of the State, its power and weakness. Statistics must chiefly aim at a true conception of the affairs of the State, so as to make the student of this system fit for public service.   Statistics deals with the *present* time, not with the past; the history of changes in the State ("Staatsveränderungen") should only be given as an introduction.

A sketch of this description will easily show how little it had to do with statistics in the modern sense of the word.   In a few pages he gives the outline of the history of Spain.   Then follows a short description of the climate, the geographic position, division of the realm and its produce; the colonies are enumerated, with a few words concerning their production and other facts.   Next follows, without any critical remarks as to the method of calculation, an estimate by Uztariz of the number of inhabitants based on an enumeration of taxpaying families, further reflexions on the causes of the sparseness of population, a description of the national character of the Spaniards, an outline of the Constitution, with remarks regarding the Inquisition, the state of science, the state of industry and manufacture ("Fleiss und Manufakturen") and the Spanish commerce, but with hardly any numerical facts.   Finally, there is Currency and Finance, also with very few numerical observations, and a short sketch of the military and naval forces, with a concluding paragraph regarding the "interest" of Spain, viz. a discussion on a single page of what had been detrimental to the welfare of the country and what in the future might be likely to promote it.

After this monograph Achenwall gives similar descrip-

---

[1] *Staatsverfassung der heutigen vornehmsten Europäischen Reiche,* 5 ed., 1768, pp. 3 *sq.* and p. 39.

tions—in the edition here quoted—of seven European States. By that time, investigations regarding the number of inhabitants and other questions of vital statistics had already been made in some of the States, such as Great Britain and Sweden, but even here only the probable population was given, without any mention of the statistical material at hand. In that respect Achenwall follows entirely in the wake of his predecessors from the sixteenth or seventeenth century. Political arithmetic seems to be quite outside the scope of statistics as he conceives this system.

Achenwall's work seems to have enjoyed a great reputation. He himself modestly acknowledged what he owed to his predecessors, particularly to Conring, whom he mentions as the father of this system.[1] Several of his contemporaries, however, assigned this title to himself.

6. Achenwall's successors seemed unable to produce a real change in the system. One of the best known among them is Schlözer (1735–1809), who held a chair at the University of Göttingen after the death of Achenwall. Like his colleagues, he tried in vain to give an absolutely clear definition of the term "statistics," entangled as he was in the nebulous conception of the "Staatsmerkwürdigkeiten" in Achenwall's work. Schlözer makes the confession that it is only a question of relativity; things which at one time are important may later on be quite insignificant.[2] Moreover, statistical observations in the modern sense of the word were still extremely scanty in these publications, even where figures were easily accessible. Meusel, for instance, who in 1804 published a book on statistics (*Lehrbuch der Statistik*), devotes only five lines to the vital statistics of Sweden. Schlözer indeed recommends the use of exact figures instead of vague expressions such as "flourishing manufactures," but he was not very successful in this respect and expressions of the kind are found abundantly in books on "Staatenkunde" of this period.

[1] John, l.c., p. 7.
[2] Schlözer, *Theorie der Statistik, erstes Heft. Einleitung*, 1804, pp. 47, 53.

As remarked above, it must be admitted that information concerning finance, commerce and population was difficult to get in those absolutistic days, particularly, perhaps, in Germany. If such data existed at all, they were as a rule safely deposited in the archives of the Government, and the university professors had few opportunities of access to them. It may sound as a paradox that our generation in many cases knows much more about vital and commercial statistics in the seventeenth and eighteenth centuries than our ancestors did. Even though few statistical facts concerning that period may be dug out from the archives, we possess good additional material in our statistical knowledge about our own age, which allows us to form out of the fragments from the past a complete impression which our forefathers had no chance of getting.

Characteristic of the system was its decidedly practical character, and this naturally hindered scientific progress. As in Seckendorff's time, the aim was to give such information as might be of use in public service. Sometimes the authors were aware that this standpoint was not altogether compatible with the claims of science, so that if Achenwall was right in confining statistics to facts from the present moment only, any statistical work would be mere *waste-paper* soon after its publication. If, then, the statisticians really wished to see their books registered among scientific works, then it would be necessary to give more than an exact picture of the present time. Sometimes the authors of statistical works admitted this themselves. Thus Schlözer agreed that a statistician might describe past as well as present times. He might, as he expressed it, "let history stand still" (*Theorie*, pp. 86–7). After this concession it might have been expected that he would have conceived the object of statistics as an explanation of present conditions judged by the past, and further that the statisticians would try to show the mutual influence of the various branches of political and economic affairs. Schlözer, however, did not venture to go so far.

His solution was, in fact, curiously devoid of a real scientific mind. According to him, statistical works should contain facts, but he advises the statistician to make his lectures less "dry" by adding "history, causes and effect," thus giving them "life and interest." How old-fashioned he was will be seen from his attempt at systemising statistics with the help of a formula, as Conring had tried. This formula is in fact nothing more than a rubric of the contents. It consisted of the three words: "Vires unitæ agunt" (l.c., 159). "Vires," the fundamental forces, would here signify the population, the country and its produce, as well as the circulating money. The constitution is the "union" of these forces, whereas the government and the administration represent the third term in the formula, they are active ("agunt"). More than anything else, such old-fashioned formulæ show how deep the gulf is which separates this system from modern thinking.

Still, gradually a certain change was unavoidable. Inaccessible as the archives seemed to be, a stream of numerical observations became gradually general knowledge, and naturally such facts claimed mention in statistical works. Various journals brought facts from all parts of the world, some, merely statistical in the old sense of the word, others numerical observations.

A short enumeration of the mixed contents of a statistical "Magazine" from 1792–3 will illustrate this.[1] The reader will here find descriptions of travels, information with regard to moral, political and physical conditions in the Sardinian States, an article on some student riots, the plan of an assurance society in Berlin; further, the number of inhabitants in Württemberg, births and deaths in the Papal States, births, deaths and marriages in the Archbishopric of Osnabrück, a list of the districts of Baden, a statistical geographical description of Mecklenburg, etc.

7. A curious phase in this evolution was the so-called

---

[1] Fr. L. Brunn, *Magazin zur nähern Kenntniss des physischen und politischen Zustandes von Europa und dessen auswärtigen Kolonien.*

"Tabellenstatistik." When a description of a State followed a definite scheme, with the same rubrics throughout, then it was quite natural to try to get a bird's-eye view of all these facts by a typographical arrangement. For an easy comparison, the facts from the various States might be arranged in columns, one for each country, and with horizontal sections for each subject. Thus the Danish historian and philologist Anchersen (1700–65), in his descriptive *Statuum cultiorum in tabulis* (1741), gives a synoptical table concerning fifteen European countries. Under one rubric Italy is mentioned as "Paradisus Europæ," whereas under another rubric the religion is described as "papistica."

As the "Tabellenstatistik" only consisted of a typographical arrangement of the facts, it was not necessary to give numerical observations. But quite naturally this arrangement invited the use of numbers, if they were at hand, and even though the "Tabellenstatistik" had to begin with nothing to do with statistical tables as we are accustomed to see them, a change in this direction would appear easier here than in the usual works on "Staatenkunde."

Other publications also contributed to a change of this kind. Here the German geographer Büsching (1724–93) may be quoted. He would treat the system as part of geography. Working with an extraordinary energy under conditions which were often anything but favourable, he succeeded in collecting enormous masses of details. His *Neue Erdbeschreibung*, which he began publishing in 1754, testifies sufficiently to his energy. The work is somewhat chaotic, but there is an evident progress with regard to the treatment of numerical facts. A very characteristic illustration is his effort to estimate the number of inhabitants in Portugal.[1] His source of information was the Portuguese author Caetano de Lima, who died 1757. This author had planned a geographic historic description of all European States,

[1] l.c., II, 1754, pp. 7–8.

but he only succeeded in realising his plan for Portugal.[1]
In an appendix he gives lists of the numbers of inhabitants and hearths in towns and rural districts, and he
maintains that this material is very accurate. Still, he
makes no attempt to add up the number in order to
find the total population. The idea of an attempt of
this kind does not seem to have occurred to him. But
Büsching attacks the problem, estimating the population,
including the clergy, to be two millions.

As regards other countries he sometimes met with
greater difficulties. Thus, to find the total number of
inhabitants in Germany he calculates from an estimate
for France (20 millions). Germany being rather populous, even more so than France, he estimates 24 millions
in Germany as a probable number.[2]

8. In this way there was a bridge between the
"Staatenkunde" and the Political Arithmetic, even
though several authors representing the former system
obstinately followed the old track. An author who
deserves being mentioned here is Crome (1753–1833),
who for a long time held a chair at the University of
Giessen. Among other works he published a book on
the size and population of the European countries.[3]
In a series of tables—not very skilfully arranged—he
gives the population, the area, the density of population
and the number of inhabitants which the country would
contain if 3,000 persons lived on each (German) square
mile. He is very careful in quoting his sources and in
dealing with the various methods which have been
applied in order to estimate the numbers.

Parallel to his contributions other authors cultivated
"Tabellenstatistik" in its original form, as a means to a
comparative description of the States. Both forms of
Tabellenstatistik, especially that of Crome and his
followers, are in direct opposition to Achenwall-Schlözer's

---

[1] *Geografia historica de todos os estados soberanos de Europa*, Lisboa, 1734.
1736.

[2] l.c., III, 1757, pp. 22–3.

[3] *Ueber die Grösse und Bevölkerung der sämtlichen europäischen Staaten* (1785).

disciples (the Göttingen school) and an extremely violent and bitter discussion took place in 1806.[1] The supporters of the Göttingen school described tabular statistics as a mere hotch-potch without form or method and the compilers of such material as slaves to the tabular form of presentation, who gave only the dry bones of statistics without clothing them with the flesh of descriptive reality; they had no sense for description of the national spirit, love of freedom, genius, etc., whereas the Göttingen school was proud of cultivating "one of the noblest sciences."

This dispute died out as do all conflicts of that nature. One of the most energetic supporters of the Göttingen school, A. F. Lueder (1760–1819), in his later years even condemned statistics altogether, whether in tabular form or not, using the most violent terms. Others continued as usual, though to some degree influenced by the evolution outside of this system. At last, in 1850, the well-known economist Knies (1821–98), took up the question of the position of the "Staatenkunde" among other scientific systems.[2] He assigned the name of statistics to political arithmetic, whereas he would give Achenwall's statistics the name of "Staatenkunde" or similar names ("Gegenwartskunde" or "Staatenkunde der Gegenwart"). In point of fact, this was already the result of the actual evolution, not only outside Germany but also within it, and thus this curious chapter of the history of statistics came to an end.

In conclusion, summing up the result of the whole evolution, it may perhaps be said that the system, which Conring, Achenwall and their disciples had cultivated, had in the long run been more influenced by political arithmetic than the reverse. At all events, its saga, as one of the sources of statistics, is out, and but for the curious change of names which has taken place, and which has often puzzled students of statistics, little interest would have attached to it.

---

[1] John, l.c., pp. 128 *sq.*
[2] *Die Statistik als selbstständige Wissenschaft*, Kassel, 1850.

Curiously enough, a new change of name was made later on, the expression Political Arithmetic sometimes being used by authors to denote simple or compound interest, annuities, lotteries, life-insurance, the theory of Calculus of probability, etc.[1]

[1] We meet this already in a work from 1782–4 by J. A. C. Michelsen, *Anleitung zur juristischen, politischen und ökonomischen Rechenkunst;* later on we may quote Wild, *Politische Rechnungs Wissenschaft. Anleitung zu allen im Staatsleben vorkommenden Berechnungen durch Beispiele erläutert*, München, 1862; Joseph Haberl, *Lehrbuch der politischen Arithmetik enthaltend die Wahrscheinlichkeits-Rechnung, die Zinzeszinsen-Rechnung, die Rückzahlungsformen von Anlehen, die Construction von Tilgungsplänen, die Lotterie-Anlehen und die Berechnung der in das Gebiet der Lebensversicherung gehörigen Fälle*, Wien, 1875; Cantor, *Politische Arithmetik oder die Arithmetik des täglichen Lebens*, Leipzig, 1898; Wilhelm Ludwig, *Lehrbuch der Politischen Arithmetik*, Wien und Leipzig, 1907.

CHAPTER III

POLITICAL ARITHMETIC IN THE SEVENTEENTH CENTURY

9. THE cradle of political arithmetic stood in London, where in 1662 a merchant, John Graunt (1620–74), published a remarkable book, *Natural and Political Observations upon the Bills of Mortality*.

At that time London had reached a large size, with some hundred thousand inhabitants. The inconvenience in those times of such crowding is evident; even in much smaller communities serious troubles were felt. To get the regular daily supply of the necessities of life was difficult enough, but still worse were the constantly threatening dangers to public health. The history of London shows a considerable number of plague-years. In 1348 the Black Death reached London, then a place with perhaps from 30,000 to 40,000 inhabitants. In 1563 there was an outbreak of plague, and again in 1592, and the seventeenth century had several severe plague-years, as in 1603, 1625 and 1665, in which a large percentage of the population died. No wonder that the citizens followed public announcements concerning burials and christenings with very great interest. Lists of births and deaths had been introduced already in the sixteenth century, and weekly reports were published every now and then, particularly in the latter end of the century. After the great plague in 1603 these weekly publications became regular, with a summary on the Thursday before Christmas for the whole preceding year. The lists contained the numbers of christenings as well as of burials, with special information about deaths from the plague. The causes of deaths, reported by sworn searchers ("ancient matrons" as Graunt calls them), appointed in each parish, were published regularly from 1629. The nomenclature was of course very incomplete, and the searchers, who had to see the deceased person before the burial, were undoubtedly often unable to give a correct diagnosis. Some diseases were even frequently left out of consideration. Graunt especially distrusts the reports as to the number of deaths from "french pox" (syphilis), and he is also well aware of other defects, for instance, as to the number of christenings, Catholics and Nonconformists buried in their own churchyards (perhaps 5 per cent. of the total number) who were not included in the lists of deaths.

The material was thus anything but good. Still, it was sufficient for Graunt to open up quite a new field for scientific investigation, and incomplete as were his observations, imperfect as his own conclusions, he was able to throw light upon subjects hitherto unnoticed. How little was known at this time may be seen from a remark by Graunt when discussing estimates of the number of inhabitants in London: "one of eminent Reputation was upon occasion asserting, that there was in the year 1661 two Millions of People more than *Anno* 1625 before the great Plague." [1] In fact, it was at the beginning of the nineteenth century that London first reached one million inhabitants. Of course there were several ill-founded popular ideas with regard to life and death, for instance, that "great *Plagues* come in with *King's* Reigns because it hapned so twice viz *Anno* 1603 and 1625" (l.c., p. 369), not to speak of the influence upon health of the phases of the moon. One of the conceptions of those days which Graunt mentions is that there were three women to every man.

John Graunt was born in 1620, the son of a draper in London. He was himself apprenticed to a haberdasher and followed this trade during his whole life.

---

[1] Quoted from a reprint of the fifth edition in Ch. H. Hull, *The Economic Writings of Sir William Petty*, Cambridge, 1899, p. 383.

Gradually he acquired some learning. He studied Latin and French in the morning before starting his day's work. Unfortunately, he had no mathematical training, which would have been very useful for his study of the Bills of Mortality. He enjoyed much reputation among his contemporaries and had several friends in the academic world. One of these friends was William Petty (1623–87), the inventor of the name Political Arithmetic. Graunt assisted Petty in his University career, and their names are so closely connected, that Petty has sometimes erroneously been considered the author of Graunt's book.[1]

The "Observations" soon secured Graunt a name in the world of science, and he was at once elected a member of the newly-founded Royal Society. He did not, however, take much part in the proceedings of this learned society. The book which brought him world-fame was in fact the only one he published.[2]

10. Quite naturally Graunt was struck by the regularity of certain statistical phenomena. This is one of the leading ideas in his book. He found that "among the several *Casualties* some bear a constant proportion unto the whole number of *Burials*" (such as chronic diseases, accidents and suicides) whereas "*Epidemical* and *Malignant* Diseases, as the Plague . . . do not keep that equality." These results are in themselves an achievement which deserved notice. No wonder that he was tempted to overrate the importance of this regularity; it is a mistake he shares with many of his successors in political arithmetic. This conception of the regularity of statistical phenomena leads him to draw conclusions as to vital statistics in the whole country outside London, from a list of christenings and

---

[1] For instance, by the Marquis of Lansdowne in *The Petty Papers: Some unpublished papers of Sir William Petty*, I–II, London, 1927; compare M. Greenwood, "Graunt and Petty," *Journ. Stat. Soc.*, XCI, 1928.

[2] The first edition appeared in 1662; the fifth edition was published in 1676, two years after the death of the author. But these editions differ very little from each other, save for some appendices, for instance, tables regarding Tiverton and Cranbrook.

burials in a single country parish (Petty's birthplace, Romsey). He is not to be blamed for not having seen that the regularity must naturally be less prominent where the number of observations is small. On the whole, looking at this first attempt at a survey of an entirely new field of observations there will be more inducement to admire his keen efforts than to criticise his mistakes and hasty conclusions.

It may be useful to go through his work in some detail in order to see the problems with which a political arithmetician in those days had to grapple, in addition to ascertaining which of his results might be looked upon as relatively safe and which of them it would be necessary to lay aside.

His "Observations" seem to be the result of several years of study, and his ideas with regard to the various problems took shape gradually. No wonder, therefore, that his results are not always in inner harmony, though his conclusions generally testify to his natural common sense.

Graunt opens his work with a description of the Bills in the very summary form in which they appeared up to the year 1629, when the totals of christenings and burials were divided according to sex, and the causes of death were reported in alphabetic order (in the year 1632 altogether 63 causes). As mentioned above, the nomenclature cannot of course satisfy a modern medical statistician, even though the lists in many respects throw light upon the sanitary condition of an urban population in those days. The list for 1632, embracing 9,535 deaths, commences with 415 "abortive and still-born," 1 "affrighted" and 628 "aged." As to the latter group, Graunt has various hypotheses: first, that the number of years which the searchers called aged "must be the same that *David* calls so, viz 70" (p. 352); second, that the persons concerned were above 60 years old (p. 348). 2,268 cases were registered as "chrisomes" (newly baptised) and infants, 267 died from "Dropsie and Swelling," 1,108 from fever, 9 from "Scurvy and

Itch," etc. Graunt makes critical remarks as to various causes. He found thus that "rickets" were not mentioned before 1634, whereas "livergrown" before that year represented this cause of death. He tries to prove (p. 365) that the numbers of deaths from the plague returned in the lists ought to be considerably increased. For in the years before and after the plague-year 1625, the ordinary number of burials was between 7,000 and 8,000, whereas in 1625 the number of deaths from the plague was returned as 35,417, and of all other diseases 18,848, so that probably 11,000 of the latter were due to the plague.

He now adds up the registered cases of deaths for the twenty years 1629–36 and 1647–58, getting a total of 229,250, out of which about 16,000 are from the plague. Leaving out the latter, he tries to find how many of the remaining died before 6 years of age. He enumerates (p. 349), not quite exactly, 71,124 deaths from various infantile causes (thrush, convulsions, rickets, teething and worms, etc.), among which are 8,559 "Abortive and Stillborn," and, moreover, he finds 12,210 deaths from small-pox, measles and other diseases which might attack adults as well as children. Assuming that about half of these were under 6 years of age, he finally finds "that about thirty six *per Centum* of all quick conceptions died before six years old." This result is in comparative harmony with later observations.

Graunt proceeds to the number of christenings. He observed that a great change has taken place, especially in 1648, "when the differences as in Religion has changed the Government" (p. 361). By comparing the stillborn and the deaths in childbed in 1631 and 1659 he draws the conclusion that the true number of christenings in 1659 was double the number registered.

By studying the number of christenings he further finds that the depopulation of a place like London by reason of the plague will be repaired in a couple of years (p. 368).

Very interesting are the efforts which Graunt makes

in order to find the *number of inhabitants* (pp. 383 *sq.*). He assumes that the normal number of christenings in the years in which the "Registries were well kept" is 12,000. This number will correspond to 24,000 "Teeming-Women," the number of these being about double the births. He further assumes that the total number of families is twice the number of Teeming-Women, or 48,000, and that the average number of members of a family is eight, viz. husband and wife, three children and three servants or lodgers. The number of inhabitants will thus be 384,000. This is corroborated by an enumeration of the families in some parishes. He found here 3 deaths yearly among 11 families, and the normal number of deaths being 13,000, this would again lead to about 48,000 families. He also studies the map of London within the walls, finding as a probable estimate that there are nearly 12,000 families. But outside the walls the number of deaths is three times the number within this space, which again leads to the conclusion that the total number of families is 48,000.

To a modern statistician these conclusions may appear rather bold. Still, they were perhaps not very far from the truth. A rate of mortality of 3 out of 88, the total supposed number of members of 11 families is not improbable, viz. 34 per thousand. Graunt's successors in the following century would be more inclined to use a short-cut, based on the crude rate of mortality or on the birth-rate, without using the number of families as an intermediate link, 13,000 deaths giving, for instance, if the death-rate is 34 per mille, a population of about 380,000.

11. Less convincing were Graunt's efforts to find the distribution of the population according to *sex* and *age*. Whereas his results regarding the mortality of children under 6 years of age, as remarked above, are not improbable, he had no guide as to the age-distribution of deaths above 6 years. He quite arbitrarily allots 24 per cent. of all deaths to the ages between 6 and 16, which is evidently a very great exaggeration. In the following

decades the estimated numbers are 15 per cent., 9 per cent., etc.   He concludes that out of 100 children 64 will remain alive at 6 years, 40 at 16, 25 at 26 years, etc., a conclusion which requires us to ignore migrations, whereas there is a balance between births and deaths, the population being stationary.

Graunt's result will appear from the following table:

Age (Years).			Number of deaths per cent.	Age (Years).			Number Surviving.
0–6[1]	.	.	36	6	.	.	64
6–16	.	.	24	16	.	.	40
16–26	.	.	15	26	.	.	25
26–36	.	.	9	36	.	.	16
36–46	.	.	6	46	.	.	10
46–56	.	.	4	56	.	.	6
56–66	.	.	3	66	.	.	3
66–76	.	.	2	76	.	.	1
76–86	.	.	1	86	.	.	0

After these numbers, between 6 and 56, about three-eighths of the persons surviving at each age will die in the course of the following decade.   According to modern experience only a small percentage will die in the young ages.   According to the experience for Greater London, 1920–2,[2] the number of deaths between 6 and 16 in both sexes combined would be only 2–3 per cent. instead of $37\frac{1}{2}$, and whereas according to Graunt three-quarters of the number surviving at 6 would die within the following thirty years, the new experience finds only 10 per cent. for males and 8 per cent. for females. Graunt supposed that the rates of mortality were approximately independent of age through a long period of life.   Even if the risk of life differed much in those times from what we are accustomed to to-day, this hypothesis could not in the long run be maintained.

But in spite of these defects we cannot help admiring

[1] It may be a little doubtful whether Graunt meant death under the completed sixth year or only in the first five years, etc.   The former alternative seems most probable.

[2] The Registrar-General's Decennial Supplement, 1921, Part I, Life Tables, 1927.

Graunt for this first attempt at finding the law of mortality. It is true that his table was very imperfect, but this means little considering that he had opened a very important field for scientific investigations.

Graunt now proceeds to the calculation of the age distribution of the population. Here his usual common sense seems to fail him. He wants to know how many "fighting men" there are in London, viz. males between 16 and 56. In order to find this number he subtracts the number surviving at 56, or 6, from the number surviving at 16, or 40. He thus finds 24, and he considers this the relative number at the given period of age, out of a total population of 100. He does not see that he has found the number of *deaths* between the two ages, instead of the number of *living*, or of persons exposed to death. Using an interpolation of the first degree, the result of which does not in fact differ much from the results of more refined methods, we can proceed in the following way. Between 26 and 36 the numbers exposed to risk will be about ten times the average of the numbers of those surviving at the beginning and the end of this age period, viz. 25 and 16. We thus find 205. For all ages together the total number would be 1,822 for each 100 children beginning life, and between 16 and 56 alone it will be 740 or 41 per cent. If he had followed this method, he might have been led to the conclusion that the number of inhabitants could not reach 384,000; if the normal yearly number of deaths is 13,000 he would find 237,000 inhabitants, and this number being remarkably low he might have been induced to revise his calculations.

As to the relative number of *males* and *females*, Graunt gives some interesting results (pp. 374 *sq.*) which are now familiar to all statisticians but which were at that time new and even surprising, viz. that there is almost balance at birth between the two sexes, with a small excess of boys over girls. For a long series of years he finds that 139,782 males were christened as against 130,866 females. For the sake of calculation, making

them 140,000 and 130,000 respectively, he finds that
there are 14 boys to 13 girls.   (This would correspond
to 519 boys out of 1,000 christenings, whereas the exact
number was 516.)   For Romsey he finds 3,256 male
christenings to 3,083 females (not absolutely correct
according to his tables).   Adjusting the numbers in
the same way to 3,200 and 3,000, he finds the proportion 16 to 15.   He seems to suppose that the number
of deaths will be proportionately the same in the two
sexes, so that there will be the same proportion between
males and females alive as between them at the time of
christening.   He does not, however, go much into the
experience regarding the sex proportion at death, and
his remarks concerning the mortality of the two sexes
are not quite clear.

As to the increase of the population, he finds that the
number of inhabitants in the country (viz. in Romsey)
only increased by one-seventh in forty years (p. 371).
In the index he draws from this result the conclusion
that the population will double in 280 years, a mistake
which other political arithmeticians have also made.   A
little more than two centuries would have sufficed (208
years).   He also has curious reflexions with regard to
the growth of London through excess of births over
deaths.   Supposing that the 24,000 "pair of Breeders"
in 8 years begot altogether 96,000 children, this will
give a doubling of future "breeders" in 8 years, but as
the total population is 8 times the breeders, a doubling
of the number of inhabitants will require 64 years.   His
reasoning is, however, not quite clear (compare Hull,
l.c., 388, the note).

These problems are not the only ones which Graunt
touches.   Thus he tries to show from his London observations that in "sickly Years" the birth-rate decreases
(p. 368).   By inspection of the material it will be obvious
that the difference is so small that nothing can be concluded with certainty.   Nor will his conclusions (p. 390)
in this respect regarding the frequently mentioned country
parish hold good.

Having thus reviewed the contents of Graunt's work, we may be justified in saying that this first attempt at vital statistics in spite of its evident defects ranks very high on account of its initiative and its freshness of view. Graunt was stimulated by his scanty and imperfect observations to attack a number of important problems, and even though he did not master them all, even though several of his conclusions invite objections, we cannot deny him the honour of having initiated a new science through his "Observations."

Graunt's mortality table quite naturally found supporters in other countries. Thus in France in 1706 M. des Billettes, member of the French Academy, proposed a tontine on the supposition that the members would die in accordance with his table, which he asserts is based on "a long experience."[1] We shall even see that the table was quoted as late as 1797 as illustrative of the mortality of mankind.[2]

12. In Holland also we meet Graunt's influence. In this country life annuities ("gewone lijfrenten") and tontines were frequently used as a commodious form of public debts. As a rule, a life annuity cost the same, whether the annuitant was young or old. If the rate of interest was 4 per cent. the annuitant would get 8 per cent. yearly ("$12\frac{1}{2}$ years' purchase"). In the tontine the income of the funds was distributed between the members till at last all members were dead. Thus in the city of Kampen in the year 1670 a public loan of 100,000 fl. was arranged with 400 shares at 250 fl. each. Jacob van Dael ("geswooren Makelaer der Stadt Amsterdam") undertook the arrangement and wrote a pamphlet in order to explain the plan and describe the advantages of the tontine compared with common life annuities.[3] After the death of all the members the city was to be the inheritor of the funds. Van Dael

---

[1] *Correspondance des Contrôleurs Généraux des Finances avec les Intendants des Provinces*, Publiée par A. M. de Boislisle, II, 1883, p. 570.

[2] Hufeland, *Die Kunst das menschliche Leben zu verlängern*, p. 214.

[3] *Bouwstoffen voor de Geschiedenis van de Levensverzekeringen en Lijfrenten in Nederland*, Amsterdam, 1897, pp. 264 *sq.*

tries to show how 400 members of the tontine, entering at one year old, would gradually die out. The scale very much reminds one of Graunt's table, only the decrease is still more rapid. Out of 400 at one year of age only 200 are supposed to survive at 12. Consequently each member alive will now get 8 per cent. for a year, or 20 fl. In the following 12 years the members will be reduced to 100, and again from 24 'till 36 there will be a reduction to 50, etc., leaving 12 at the age of 60. In the following years the decrease will be still more rapid; at the age of 72 only 3 would be alive, and at 80 none would be left. Whereas according to Graunt's hypothesis the probability of death within a year during a long period of life was 4–5 per cent., here we should find 5–6 per cent. Van Dael himself maintains that his table is the result of accurate observations.

Still more evident is the influence of Graunt on some investigations undertaken in 1669 by the famous mathematician Christiaan Huygens (1629–95) and his younger brother Lodewijk, with whom he corresponded. The latter quotes Graunt's mortality table which apparently he considers trustworthy. He tries to find the mean duration of life and arrives at the same result as above, viz. 18·22 years of life for a new-born. Curiously enough, he seems to start with the moment at which pregnancy of the mother begins. His elder brother criticises the result, referring to his˙ investigations of the calculus of probabilities which will be mentioned below. He maintains that the mean duration of life conceals great differences, some persons dying early, others attaining a high age, and he wishes to know the probability of survival or death at a given age. He agrees with his brother in recommending the mean duration of life as a means to calculate the value of life annuities.

After having seen some observations made by Johannes Hudde (1628–1704) in 1671, on persons on whose lives annuities had been bought in the years 1586–90,

Chr. Huygens felt less sure with regard to Graunt's table. Hudde's observations might easily be used to form a life-table,[1] but he did not undertake such calculations. The result is a table with a much more favourable mortality than that according to Graunt or van Dael. On the other hand the chances of death nowadays are much smaller, even when compared with Hudde's experience for periods without plague-years.

About the same time the famous politician Johan de Witt (1625–72) wrote a remarkable paper on the value of life annuities.[2] In this paper he refers to observations concerning several thousand annuitants, and he gives a law of mortality without, however, explaining how he found it. From the fourth year of life, according to his table, 128 persons are gradually reduced, one dying every half-year during 50 years, so that 28 survive. In the next 10 years one death will take place every 9 months, so that $14\frac{2}{3}$ are left in the sixty-fourth year. In the following decade one-half will die in each six months, so that $4\frac{2}{3}$ will survive, and finally one-third would die every half-year till after 7 years all have died out. It seems difficult to see how he arrived at this result. He cannot have confined himself to the observations which Hudde had collected, for this material is much less than the number of observations to which he refers. The rates of mortality according to his table are higher than those according to Hudde's observations, but his table gives at all events a better picture of mortality in its dependence on age than Graunt's and van Dael's tables, even though he seems to have used the observations at his disposal very freely.

The clearness of his thought appears from the fact

---

[1] *Bouwstoffen*, pp. 85 *sq.*; confer Westergaard, *Die Lehre von der Mortalität*, 2nd ed., 1901, pp. 270 *sq.*

[2] *Waardije van Lijfrenten naar proportie van Losrenten* (1671). Confer Fr. Hendricks' *Contributions to the History of Insurance and of the Theory of Life Contingencies*, London, 1851; Deuchar, "A Sketch of the History of the Science of Life Contingencies," *Transactions of the Insurance and Actuarial Society of Glasgow*, 1882; G. Eneström, *Ett bidrag till Mortalitätstabellernas historia före Halley. Ofversigt af Kongl. Vetenskaps-Akademiens Förhandlingar*, 1896.

that he calculates the value of a life annuity quite correctly instead of using the mean duration of life, as Huygens proposed. But the calculation was more laborious than at present, as he was unaware of the technical process now used.

Interesting as these contributions to vital statistics are, it was not till the eighteenth century that Holland reached the position of one of the leading countries with regard to Political Arithmetic.

13. After this digression we may return to England, the first name met here being that of Graunt's friend, William Petty. He was a very talented man who from humble conditions gradually rose to a high social position. At one time he was professor at Oxford University; later he was appointed physician to the Army in Ireland, where he undertook to survey the land forfeited on account of the rebellion of 1641. In that way he founded his own large fortune. Though a firm supporter of the family of Cromwell, he won the favour of Charles II and was knighted by him in 1661. He was one of the founders of the Royal Society.

In his writing, Petty shows great power of imagination, and at the same time he takes great interest in practical political questions. In his *Political Arithmetick*, which was probably written in 1671–6, but not published till 1690, he tries to prove: "That *France* cannot, by reason of Natural and Perpetual Impediments, be more powerful at Sea than the *English* or *Hollanders*" (Hull, l.c., p. 247); in another essay he maintains that "London hath more *People, Housing, Shipping* and *Wealth* than *Paris* and *Rouen* put together" (l.c., p. 509). The problems of vital statistics which Graunt had discussed did not seem to interest him. He takes it for granted, on Graunt's authority, without entering into detail, that the rate of mortality in London is $\frac{1}{30}$ (l.c., p. 459). But he is keenly interested in the question of numbers of inhabitants, and here his powers of imagination found beautiful play. As he wrote essays on political arithmetic so he also wrote on *The Political Anatomy of Ireland* (written

about 1672, published 1691). Here he asserts that
Ireland in 1641 had 1,200,000 inhabitants (l.c., p. 154),
but 500 years earlier only 300,000; the population might
double itself in 200 years, but on account of epidemic
diseases, famines, wars, etc., he assumes that doubling of
the population will require 250 years.   In another essay
he enters into the problem of the population of London
(*Five Essays in Political Arithmetick*, published 1687, l.c.,
pp. 533 *sq.*).   He first tries to find the number of houses
in London.   In the Great Fire of 1666, 13,200 houses
were burnt, representing one-fifth part of the whole, which
he concludes from the proportion of the deaths in the
destroyed part of London to the total number of deaths.
Consequently in 1666 the number of houses would be
66,000; but taking the proportion of deaths in 1666 to
1686, we find an increase of from 3 to 4, so that we may
infer that the total number of houses in 1686 was 88,000.
This he finds corroborated by a study of the map of
London, giving 84,000 houses in 1682.   As London
doubled in 40 years, the number in 1686 would probably be
10 per cent. more than in 1682, or 92,400.   He does not,
however, base his calculation of the number of inhabitants
on these results.   Finding that Dublin's 6,400 houses had
29,325 hearths, London's 388,000 hearths would (not
quite exactly calculated), according to the same proportion, have 87,000 houses.   Corresponding observations
for Bristol would give 123,000; he then takes the mean
between these results and finds 105,000, nearly the same
number which the Hearth Office has given (105,315).
Supposing that each family consisted on an average of six
heads, and that 10 per cent. of the houses had two families
in each, the other ones only one family, he concludes that
in the calculated 105,000 houses the number of inhabitants was 695,000.   About the same result will be found
by assuming the mortality in a normal year to be one out
of 30, and, taking the average of deaths in the two years
1684–5, viz. 23,212, he finds 696,360.   Finally, he bases
a calculation on the supposition that one-fifth of the
population dies in plague-years.   Here, again, he quotes

Graunt as his authority, though Graunt does not seem to have mentioned this proportion in his book. The number of deaths from the plague in 1665 being 98,000, the number of inhabitants would in that year be 490,000, and 20 years later probably one-third more, or 653,000. It will be seen that these calculations are based on so many suppositions that it would be possible by an arbitrary change in one of the links to alter the final result considerably. For instance, Graunt supposes the number of members of a family to be on an average 8, whereas Petty says 6.

In another essay he surveys the problem of the doubling of the population (l.c., pp. 460 *sq.*). In the country mortality is one out of 50 per annum, and there are 24 births to 23 burials. This, he supposes, will give a doubling in 1,200 years. Again—assuming a rate of mortality of $\frac{1}{30}$ and a birth-rate of $\frac{5}{4} \cdot \frac{1}{30}$—he finds a yearly surplus of one out of 120, which he supposes will lead to a doubling in 120 years. Actually a yearly increase of this size would double the population in 83 years, instead of 120.

In order to find a single period of doubling he now takes various averages. When in the country one dies out of 50, in London one out of 30, he takes $\frac{1}{40}$ as normal for a whole population. Likewise he finds for the births $\frac{10}{9}$ of the deaths, which is a somewhat too low average of $\frac{24}{23}$ and $\frac{5}{4}$. The natural increase thus being $\frac{1}{360}$ yearly, he finds that the period of doubling will be 360 years (using the formula of compound interest, we should find about 250 years). As the population of London was supposed to double itself in 40 years, there would be about 5 millions in 1800, with a population outside of London of 4 millions. At this moment the growth of London must come to an end, as there would not be the necessary number of men to perform the tillage and other rural work. It could, of course, not have been expected that Petty should realise how intricate the problem was, that, for instance, the immigration into London from the country on the given suppositions as to mortality and birth-rate probably would be variable during the period concerned, and that the suppositions themselves, particularly the low birth-rate in the country, were improbable.

Petty inserts the remark that a doubling in ten years would be physiologically possible. As is well known, this remark played a part in the discussion on population which Malthus opened more than a century later.

Petty also tries to find how world population may have increased since the Flood, assuming at first a doubling in ten years, whereas later on a doubling would require more and more years. At Moses' time the number of inhabitants on the whole earth should be 16 millions, in his own time he supposed it to be 360 millions.

Petty was not the only one whose power of imagination tempted to calculations of this kind. Süssmilch mentions a curious case in his *Göttliche Ordnung*.[1] In his Vienna gloriosa (1703), Reifenstuhl, after having estimated the population of Vienna at 600,000, compares the circumference of that place (20,830 paces) with that of Rome (20,000), and concludes that Rome is less populous. Only the old Babylon, with a circumference of 45,615 paces, may have been more populous. Since the reign of the Emperor Ferdinand II (1619–37), the population of Vienna had increased from 80,000 to 600,000. A corresponding increase in the following 70 years would give 3 9 millions, and the circumference of Vienna would then be about 1 36,000 paces, Vienna consequently surpassing the old Babylon. These naïve calculations are evidently based on the supposition that the circumference is proportionate to the area.

Still, here is an attempt at finding a basis for calculations. Often the authors give estimates which apparently have no foundation at all. Thus when Isaac Vossius tries to find the population on the earth, Europe had, in his opinion, 30 millions, France alone five.[2] The Jesuit Riccioli had probably a much nearer guess (19 or 20 millions in France and 99 to 100 millions in Europe

---

[1] 4th ed., 1775–6, II, pp. 480 *sq.*
[2] Isaaci Vossii, *Variarum Observationum Liber*, London, 1685.

altogether).[1] These numbers are still quoted in the middle of the following century by Deslandes,[2] who tries to make a fresh estimate on the basis of these authorities. It may be added as a curious instance that Riccioli, after experience at Bologna, which is said to have a birth-rate of one out of 15 [*sic*], estimates the yearly number of births on the whole earth at between 66 and 70 millions.

14. A much more rational attempt at estimating a population was made by the famous English astronomer, Edm. Halley (1656–1742).

Since 1584 regular lists of births and deaths were kept in Breslau in Silesia, by that time already an important city, and it is even possible to trace births and deaths back to the middle of that century. A clergyman and scientist, Kaspar Neumann (1648–1715), was attracted by this material.[3] He tried to counteract certain popular superstitions concerning life and death. Thus he proved that the phases of the moon had no influence on health. This was also the case with the so-called *climacteric years*. It was generally believed that the human body had a seven-years period, and that consequently every seventh year would be critical to health. Particularly dangerous in public opinion were the ages of 49 and 63.[4] Gradually, however, the theory lost ground, though statistical evidence was not available.[5] Neumann seems to be the first who tried to enter into this problem by means of statistical observations. Less safe were his conclusions with regard to the *climacteric weeks*. He dealt here with the so-called *septenarii* and *nonarii*, infants in the first year of life, whose age in weeks was a multiple of 7 or 9. In point of fact, the rate of mortality from one week of age to another in the first year of life varies so much that it would

---

[1] *Geographiæ et Hydrographiæ Reformatæ Libri Duodecim. . . . Auctore,* Baptista Ricciolio, Bononia, 1661.

[2] Deslandes, "Conjectures sur le nombre des Hommes qui sont actuellement sur la Terre" (*Recueil de différents Traités de Physique et d'Histoire naturelle,* II, Paris, 1750).

[3] J. Graetzer, *Edmund Halley und Caspar Neumann,* Breslau, 1883.

[4] Cp., for instance, Stechanius, *De Annis Climactericis,* Erfurt, 1633.

[5] Thus, for instance, Schwartzwaldt, *Disputatio Physica de Annis Climactericis Vitæ Humanæ,* Wittenberg, 1682.

require better methods than Neumann's to get sound results.

These investigations were based on observations for 1687 and the following time, and Neumann's results were communicated to his great contemporary, Leibnitz, who seems to have sent them to the Royal Society in London. On the whole, the learned societies of those days had rather a lively correspondence with each other. The society forwarded Neumann's investigations to Halley, asking for his opinion, and this led to a correspondence which enabled Halley to get the observations in the form which he found necessary for his purpose. The result was his famous paper, based on the material for the quinquennium 1687–91: "An Estimate of the Degrees of the Mortality of Mankind drawn from curious Tables of the Births and Funerals at the City of Breslaw."[1]

Halley's contributions to vital statistics can hardly be valued too highly.[2] With a wonderful clearness of thought he gave the problems before him the right form. Unfortunately his successors did not always fully understand him, so that his investigations failed to command the influence which they deserved.

Halley works with the available material under the supposition that the population is nearly stationary. The lists at his disposal embracing the Lutheran inhabitants of the city gave as a main result for the quinquennium, 5,869 deaths and 6,193 births, leaving an average annual surplus of 64. This difference may be balanced by the levies for the Emperor's service in his wars.

He then proceeds to calculate the probable number of inhabitants in Breslau. In an average year, with about 1,238 births, he finds 348 deaths under one year of age, thus 890 would be alive on their birthday one year old. But the problem is: how many children under one year

---

[1] The *Philosophical Transactions of the Royal Society of London* for 1693, with a supplement in the same volume, "Some further considerations on the Breslaw Bills of Mortality."

[2] Cp. R. Böckh, "Halley als Statistiker" (*Bulletin de l'Institut international de statistique*, Tome VII, 1893, and Westergaard, *Die Lehre von der Mortalität und Morbilität* (2nd ed., 1901, pp. 34 *sq.*).

of age would be found alive at a certain moment, for instance, on New Year's Day. In order to find this number particular observations were required.

For this Halley had sufficient data at his disposal, at least for the year 1691. Out of 1,218 children born in this year, 226 died before New Year, thus leaving 992 alive. Quite naturally Halley adjusted the result to 1,000, so that a census at New Year would have given 1,000 children under one year of age. Sometimes, however, this has caused a little confusion, Halley's calculations being misunderstood, as if 1,000 were the number of newly born; Dan Bernoulli even, as will be mentioned later, supposed that Halley meant the number of children who had just reached their first-year birthday.

Having now found the population under one year of age, Halley proceeds to the following ages. Here the distinction between the deaths according to age and to birth-year is less important, and on the whole he uses the observations somewhat freely. By some small alterations the difference between the number of births and deaths disappears. As to the ages 14 to 17 he amends the number of deaths according to experience in Christ-Church Hospital in London. In the eighty-fourth year he finds 20 persons surviving; for the following ages he gives no age distribution; he only estimates the number above 84 at 107, so as to bring the total up to the round number of 34,000. This estimate of Breslau's population at the close of the seventeenth century seems to be fairly well in harmony with other observations.

Having now constructed a table giving the population according to age, Halley is able to solve various problems. A chief problem is to find the number of "fencible men." About 18,000 out of 34,000 will be between 18 and 56, and at least 9,000 of these will be males. And upon the supposition of balance between births and deaths he can further show how a generation of 1,000 persons between 0 and 1 will gradually die out. Thus, for instance, he will be able to see at which age, "it is an even Lay, that a Person of any Age shall die" (viz. that just half of a

certain number of persons of a given age will be surviving).
But the most far-reaching conclusions pertain to life
insurance. Thus, if the rate of interest is given, his table
permits him to calculate the value of annuities on one,
two or more lives (in "years' purchase"). Like Johan
de Witt, he uses a perfectly correct method. As the
technical process reducing the work considerably is of
a much later date, he naturally finds the calculations
"most laborious."

15. It is not easy to say how far Halley's table gave a
correct picture of the population and its mortality.
During the seventeenth century the number of inhabitants
was not, it seems, very fluctuating,[1] apart from the plague
years, such as 1633 with its 13,231 deaths, and we may
perhaps suppose, as Graunt did with regard to London,
that a depopulation on account of the plague was soon
remedied. We may further suppose that there was
balance between births, deaths and migrations, so that
the loss by deaths and surplus emigration was covered
by the births. If we consider the years 1687–91 as
normal we are entitled to look upon the total number of
inhabitants, 34,000, which Halley found, as tolerably
correct. Even if he had not made his ingenious distinction as to birth-year and age at death, his result would
probably not have deviated much from truth. If he had
considered all the deaths under one year as taking place
in the birth-year, the grand total would have been
estimated a little too small, but the error would not have
reached 1 per cent. But having made this distinction,
he was entitled to look on the number of children in
each year of age as approximately true.

And upon the same suppositions, the same may be
said of the remaining part of the population. But then
the question arises whether the table can at the same
time serve as a life-table. As to the five years 1687–91,
it must be remembered that there was a small difference
(5 per cent. between deaths and christenings). If the
balance is obtained by a surplus emigration, the rates of

----

[1] Graetzer, l.c., p. 94.

mortality which Halley found for this period will be slightly too large.

But it must be borne in mind that this quinquennium was uncommonly favourable with regard to health. Thus, during the years 1692–1731 mortality in Breslau seems, roughly speaking, to have been about 20 per cent. higher than according to Halley.

Halley's life-table could not, therefore, in the long run, serve as a true expression for the mortality in that place, and still less could it be accepted generally for a whole population or a select class. But at all events it came much nearer to truth than any other table which had hitherto been known. Of course, his rates of mortality are much higher than the figures which are found nowadays under modern hygienic conditions. Thus the rates for children under 7 are very high; from the eighth year they are reduced to about 1 per cent., with a minimum at 12 years; after that age an increase takes place just as in modern life-tables.

It follows from what has been remarked that it is a question whether a life-office would have benefited from Halley's table as a basis for its business. It would have been necessary to have a broad margin for eventual losses. But it is a fact that Halley's contemporaries hardly understood his calculation. At the close of the seventeenth century two life-offices were founded in England, and in 1706 a third society, The Amicable, was established, which remained in existence till 1866. Other life-offices followed, but none of them had a really rational basis, and in public opinion life insurance was to a great extent a game of hazard. Many years later, premiums were first calculated according to Halley's methods, and in 1765 a new life-office, The Equitable, was founded, with relatively rational tariffs.

In the foregoing, Leibnitz's name has been mentioned in connection with the Breslau material. He, however, only occasionally took any interest in problems of political arithmetic, but his ideas have a wide horizon, as might have been expected from this great genius. He wanted a central statistical office for the benefit of all branches of public administration, with bills of mortality, lists of baptisms and marriages; and from those observations it would be possible to draw conclusions as to population and to the Power of the State. In a paper, *Questiones calculi politici* . . . , he has a list of fifty-six questions for which he desired an answer, namely, with regard to density of population, ages distribution, marital condition, number of unmarried women, of men who are able to carry weapons, mean duration of life, causes of diseases and death, etc. In a report of 1700, in which he supports the plans of a Kingdom of Prussia, he estimates the population altogether at nearly 2 millions, the birth-rate being $\frac{1}{30}$ and the yearly number of births 65,400. He even considers this estimate as a minimum, the birth-rate being in his opinion probably somewhat smaller.[1] He may have been induced to this conclusion by his wish to get a relatively high number of inhabitants; it is more likely that the estimated population was a little smaller.

[1] Behre, *Ueber den Anteil germanischer Völker an der Entwickelung der Statistik*, Statistisches Archiv, 1907.

CHAPTER IV

AROUND THE YEAR 1700

16. As we have seen, political arithmetic had a promising start in the seventeenth century. Three authors on the threshold of the eighteenth century made weighty contributions to the new science, viz. Vauban, Davenant and Gregory King. The famous maréchal Vauban (1633–1707) wrote a remarkable book, *Projet d'une dixme royale*. It was finished in 1699, but did not appear till 1707. The fate of the book and of its author is well known. On account of Vauban's open description of the sufferings of the French people his book was publicly burnt by the executioner, and he himself fell in disgrace and died shortly after. The book contains several interesting contributions to political arithmetic, thus as to the effect of the system of taxes which he recommends, and remarkable attempts at finding the area of France by means of various maps. How great the difficulty of such attempts must naturally be in those times appears from the fact that the highest estimate of the area is 23 per cent. higher than the lowest one. He also tries to make an estimate of the number of inhabitants, and arrives at the conclusion, which was probably not far from the truth, that France contained 20 millions, much less, according to his opinion, than might live there under happier conditions. It is very interesting that he —like Leibnitz—recommends official statistics, for instance, concerning religious profession, foreigners and public buildings. But the time was not ripe for such proposals; generally it was considered impossible to get reliable information in that way. On the whole, political arithmetic was for a long time very little culti-

vated in France, till about in the middle of the century a new era began.

17. The other two authors who are to be remembered here are Gregory King (1648–1712) and Sir Charles Davenant (1656–1714).

The *former* was genealogist, engraver and herald; he held an office as "Lancaster Herald," but was for a while secretary to the Commissioners for receiving and stating the public accounts.

Davenant (or d'Avenant) was Commissioner of Excise, later Inspector-General of Exports and Imports.

Both had opportunities of getting statistical data, which they treated of in various works. They did not follow Graunt's or Halley's line, being more interested in investigations of the same character as William Petty's, and quite naturally Davenant considered Petty as the founder of political arithmetic.

English economic literature was at this time in a flourishing state. It is sufficient to mention W. Temple, Josiah Child and Dudley North. Davenant may be justly ranked between them. I shall, however, only deal with his works as far as they concern political arithmetic. King's chief work was his *Natural and Political Observations and Conclusions upon the State and Conditions of England* (1696). It was, however, not published in his lifetime, but Davenant made free use of it and quoted it largely in his *Essay upon the Probable Methods of Making a People Gainers in the Balance of Trade* (1699).[1]

Davenant made an interesting attempt at giving a theory of political arithmetic in an introduction to his

---

[1] Davenant's works are here quoted after the edition of 1771, *The Political and Commercial Works of that celebrated Writer, Charles d'Avenant, LL.D.*, collected and revised by Sir Charles Whitworth, Vols. I–V, London, 1771. King's *Observations* remained unpublished till 1801, when George Chalmers added it to his *Estimate of the Comparative Strength of Great Britain*. Chalmers also appended two tracts by King, on Gloucester (1696) and on Hospitals and Almshouses (1697). As to Davenant's economic writings see W. Caspar, *Charles Davenant, Ein Beitrag zur Kenntnis des englischen Merkantilismus*, Jena, 1930.

Discourses on the Public Revenues (1698).[1] Though ascribing the new science to William Petty, he at the same time criticises him rather severely. He defines political arithmetic as "the art of reasoning by figures, upon things, relating to government." In all Petty's "inquiries he took for guides the customs, excise, and hearth money." But to a great extent these data were incomplete, so that many of his conclusions were wrong, and his successors were misled. Charles II liked to listen to his arguments that England was on a par with France,

"but we have lately had manifest proofs, that this great genius was mistaken in all these assertions; for which reason we have ground to suspect he rather made his court, than spoke his mind. The King was well pleased to be lulled asleep by a flattering council, which suggested that the power of France was not so formidable."

But it is important for a statesman to get true knowledge of his own country compared to other countries,

"and mankind in a mass being much alike everywhere, from a true knowledge of his own country, he may be able to form an idea, which shall prove right enough, concerning any other, not very distant people."
"As for example, when the number of inhabitants in England is known, by considering the extent of the French territory, their way of living and their soil, and by comparing both places, and by other circumstances, a near guess may be made how many people France may probably contain."

This reminds one of the reasoning which led Büsching to find Germany's population by comparison with France (above, Art. 7).
Further, he explains that

"in the same manner, he that knows the income of England from trade, by contemplating the frugality and industry of the Dutch, their several sorts of commerce, the places to which they deal, and their quantity of shipping, shall be able to find out what annual profit arises to the Hollanders from their foreign traffic."

---

[1] Of the Use of Political Arithmetic in all considerations about the revenue and trade (l.c., I, pp. 127 *sq.*).

"And he who knows what taxes and impositions one country can pay, by considering the different humours of the people, their stock and wealth, their territory, their soil and trade, shall be able, by comparison to form a good conjecture, what revenues can be raised in another nation; and consequently he may make a near guess how long either Kingdom can carry on a war."

He admits that there are difficulties, but a

"great statesman, by considering all sort of men, and by contemplating the universal posture of the nation, its power, strength, trade, wealth and revenues, in any counsel he is to offer, by summing up the difficulties on either side, and by computing upon the whole, shall be able to form a sound judgment, and to give a right advice: and this is what we mean by Political Arithmetic."

I have quoted at such length, as we here get a good idea of the methods followed by the political arithmeticians. It is evident how far he is from an exhaustive theory. Davenant is unable to control his estimates, to find the limits within which the numbers are probably lying, nor do we learn in detail how the estimates are to be made, and how the result can be anything but a very rough guess.

Nor is Davenant always able to keep clear of the temptation against which Petty had to fight. Thus in *An Essay of Ways and Means of Supplying the War* (1695, l.c., I, pp. 26 *sq.*) he tries to calculate what a "Poll-money" would give. According to the books of the hearth office there were about 1,300,000 houses in England, but out of these 554,631 had only one chimney, and he supposes that most of the inhabitants in these houses are poor, thus reducing the number of houses from which poll money might be expected to 800,000. In each of these houses he supposes that the average number of inhabitants is 7, giving altogether 5,600,000 persons, out of whom one-third are supposed to be able to pay 4 shillings yearly, or altogether £373,000. Further, he assumes 80,000 to be gentlemen or others who can pay £4 yearly, and 40,000 shopkeepers and

others who can pay £2. In round figures he puts it at
800,000 yearly. Here he seems to have been misled
by his wish to arrive at a high amount, for he makes no
attempt at reconciling his estimate concerning the
population of England with King's results, which he
also considers as reliable. Whereas Davenant's estimate
would lead to a population of 9 millions, King finds
$5\frac{1}{2}$, basing this, however, on a rather low average number
of inhabitants (4 to 5) in each house.

On other occasions Davenant shows a more critical
sense. Thus when discussing the balance of trade he
is well aware that the figures are not reliable, particularly
as the exports are overstated, the observations thus giving
a wrong idea of the surplus of export over import.

18. As to questions of population, he considers King
as a great authority. Speaking of the returns concerning "the present duty on marriages, births and burials,"
he says that King

"by his general knowledge in political arithmetic has so corrected
these returns as from them to form a more distinct and regular
scheme of the inhabitants in England, than peradventure was
ever made concerning the people of any other country. There
is nothing of this kind escapes the comprehension and industry
of this gentleman." [1]

As observed above, King finds the number of inhabitants to be $5\frac{1}{2}$ millions. The normal number of births
being supposed to be 190,000 yearly and of deaths
170,000, there would be a yearly increase of 20,000,
if on an average 4,000 were not supposed to die from
the plague, whereas 6,000 died in war or at sea and
more than 1,000 going to the plantations, the supposed
average increase was thus reduced to 9,000 yearly.

Proceeding to estimates of age distribution, King
evidently comes much nearer to truth than Graunt.
By comparing with Swedish census results from the
eighteenth century, we find that the number of children
was probably somewhat exaggerated by King ; he also

---

[1] l.c., I, pp. 26 *sq.*; confer about King further, II, pp. 165 *sq.*

assumes too many persons (11 per cent.) to be over 60 years, so that relatively more persons should have been placed in the intermediate ages. But on the whole he seems to have got a fair idea of the status. He also tries to distribute his numbers according to marital condition, profession, income and consumption. The results concerning consumption lead to estimates of agricultural produce. They cannot of course claim to be looked upon as entirely reliable, but founded as they are on the author's general knowledge concerning the economic conditions of his time, they testify at all events in an interesting way to the poverty of that time compared with modern wealth. Nor can we consider the famous and frequently quoted table showing how " a defect in the harvest may raise the price of corn" (l.c., II, p. 224) as based on sufficiently trustworthy observations.

King, like Petty, tried to find the world population and to compute its growth from time to time. He supposed that about 1,400 to 1,500 years before Christ the whole world contained only 4 to 5 millions. At the beginning of the present era his computation for England was 400,000 inhabitants, in 1900 it was expected to count about $7\frac{1}{3}$ millions, and in 2,300 he assumes the number would be 11 millions.

CHAPTER V

STAGNATION OF POLITICAL ARITHMETIC

19. Compared with the promising beginning of political arithmetic, the first age of the eighteenth century was poor. New ideas did not appear and statistical observations were relatively scanty. As late as in 1746 a well-known Swedish economist, Anders Berch, wrote a treatise on political arithmetic which contains nothing more than Davenant's treatise on the same subject.[1] He estimates the population of Sweden at nearly 3 millions, but adds that comparison with England would, according to his opinion, prove that Sweden could easily give 26 millions a living.

It may be that the temptation to play with the scanty numerical observations, using imagination where facts were missing, was somewhat less prominent in the eighteenth century, the political arithmeticians in continuing the work of Petty or Davenant being more sober in their calculations, though at the same time of course less attractive to a common reader. This, however, may be registered as real progress, though of a negative kind.

What has been remarked here will hold good as regards all branches within political arithmetic—vital as well as economic statistics.

As to medical statistics, two German physicians may be mentioned here, viz. Daniel Gohl and Christian Kundmann.[2] Gohl lived in Berlin, where in 1707 he started a journal *Acta Medicorum Berolinensium*. In this

---

[1] *Sätt åt igenom Politisk Arithmetica utröna Länders och Rikens Hushåldning*, Stockholm, 1746.

[2] J. Graetzer, *Daniel Gohl und Christian Kundmann*, Breslau, 1884.

journal he published lists of deaths for the year 1720 and the following years, with distribution according to causes of death, from 1721, also according to months. There was no classification as to age, but persons dying in old age were recorded. Occasionally Gohl also takes climacteric years (49 and 63) into consideration, though he himself did not believe in their influence on mortality.

Kundmann, who lived in Breslau, started a journal in the same year as Gohl, in co-operation with two colleagues. The general programme was much wider than in the Berlin journal,[1] but as to statistics no essential extension was possible. There are comparative statistical tables for various places, particularly for Breslau and Berlin, for 1722 and the following years, with distribution as to months and to causes of death, and with several notes about the ages of the deceased. He, too, has observations on climacteric years, though too fragmentary to allow of any conclusion.

None of these authors can be said to have contributed materially to medical statistics. Nor was great progress made in England in the same period. J. Smart constantly recommended improvements in the Bills of Mortality; he wanted particulars as to the ages of the deceased.[2] A reform in this direction was made in 1728, and Smart treated the observations for the Decennium 1728–37. But still the London material seems to have been very defective. In 1729 Maitland complains that lists of births are missing in several congregations, and also the death-lists were defective, though in a smaller degree.

Although political arithmeticians were probably more cautious in their conclusions, there was still a great temptation to use imaginative numbers. Best known in this respect is perhaps Abraham de Moivre's hypothesis with regard to mortality. This eminent mathe-

---

[1] *Sammlung von Natur- und Medicin-Geschichten, wie auch hiezu gehörigen Kunst- und Litteratur-Geschichten, so sich in Schlesien und anderen Ländern begeben.*

[2] See his *Tables of Simple Interest and Discount*, 1707, and later his *Tables of Interest, Discount, Annuities, etc.*, London, 1726.

matician (1667–1754) emigrated as a youth from France to England on account of the revocation of the Edict of Nantes. His contributions to the Calculus of Probability will be mentioned below. His hypothesis as to mortality is explained in his *Annuities upon Lives* (1725). With Halley's table as starting-point he gives a simple formula which—above 12 years of age—corresponds tolerably well to Halley's figures. He supposes that out of 86 newly-born individuals, one will die yearly. De Moivre uses this formula to calculate values of annuities and he still recommends his hypothesis in the fourth edition of the book (1752). In this way Halley's table was preserved from oblivion, as de Moivre uses it in all the editions for comparison, and the hypothesis had the great advantage of greatly simplifying calculations. On the other hand, this simplification prevented de Moivre from entering more deeply into the technical problems of these calculations.

Other authors made suppositions which were less in conformity with the observations at hand. This was the case with Isaac de Graaf in Holland.[1] He operates with the "vital power" ("levenskragt") of men, which according to his opinion is constantly decreasing as age advances. It is not easy to reconstruct his calculations. The "levenskragt" seems to be the complement of some power of a fraction, the numerator of which being the age of the person concerned and the denominator the highest age which he thought attainable, viz. 92.[2] By using the fifth power he finds some correspondence with de Witt's results.

20. The tariffs of annuities on life seemed in those days to depend on practical considerations rather than on theories. In his *Tables of Interest, etc.*, from 1726, Smart explains (p. 113) that as there are very few reliable observations on the numbers of persons dying at

---

[1] Waardije van Lijfrenten, Naar proportie van Losrenten, 1729 (*Bouwstoffen*, pp. 151 *sq.*).

[2] $1 - \left(\dfrac{x}{92}\right)^n$, x being the age.

different ages, the valuation of annuities upon lives is nothing but guesswork, and he recommends to the reader as the best method he can propose "to consider with himself what number of years certain he supposes to be equal to the Life or Lives in Question, having always regard to the Age, and State of Health of the Annuitant or the Annuitants"; the value of an annuity payable during a period of this length will give the life-annuity.

In one respect, however, Smart shows a deeper insight than most of his successors. He objects (p. 111) to the settling of "a certain Sum of Money in lieu of Tithes upon the several Parsons, Vicars and Curates in the Parishes of the City of London, which were burnt by the late dreadful Fire." Instead of settling a definite sum of money, the gradual depreciation of money ought to be taken into consideration. He would prefer that "instead of a certain annual Sum of Money, a certain Pound-Rate should have been settled; as suppose 8*d*. or 9*d*. per £ upon the Rents of the Houses within each Parish." An idea such as this was in fact more natural in those days than in the present period of money economy. It is obvious that the policy-holders have on their shoulders a rather heavy risk arising from the movements of the price-level, and that somehow or other the insurance societies ought to take this risk themselves. Even after 200 years of progress the insurance societies have not solved this problem. Nor did Smart, however, enter more deeply into the question; he even makes an excuse to the reader for "this Digression."

Several years after Smart, Weyman Lee wrote a voluminous work to criticise Halley's methods of calculating life-annuities.[1] According to his opinion it was only necessary to know the probable duration of life of the person concerned; an annuity corresponding to this number of years would give the value of the life-annuity. Again, if we know this value and the rate of interest we can calculate the probable duration of life, and by

---

[1] *An Essay to ascertain the Value of Leases and of Annuities for Years and Lives and to Estimate the Chances of the Duration of Lives*, London, 1737.

using Halley's figures for various rates of interest he gets different numbers for the duration of life, and this he finds (l.c., p. 152) to be so "palpable a contradiction to common Sense that nothing can maintain the Rule by which it was produced." It does not occur to him that a contradiction of this kind would arise if his supposition as to the calculation of annuities on life was wrong. He extends his naïve calculations to annuities on two or more lives, though he modestly confesses that he does not pretend to prove his rule by a mathematical demonstration, being "not Master enough of the Common Processes of Algebra to enter into that Sort of Proof, and possibly the Thing may not be capable of it" (p. 421). Fourteen years later he repeats his argument in another treatise.[1]

An ignorance of just the same kind is displayed by Richard Hayes.[2] He renders the customary annuities allowed for £100 by some companies and corporations to the annuitant upon life, and he then proceeds to give methods for calculating annuities on two or more lives. His methods are quite simple, but unfortunately entirely wrong.[3]

21. It might have been supposed that *France* at that time had favourable conditions for political arithmetic. Already on the threshold of the seventeenth century, Sully [4] (who was appointed superintendent of finance

---

[1] *A Valuation of Annuities and Leases certain for a Single Life*, London, 1751.

[2] *A New Method for Valuing Annuities upon Lives*, 2nd ed., London, 1746.

[3] Thus to find the value of an annuity upon two lives he gives the following rule:

First, Find out how many Years' Purchase the Life and Age is valued at.

Secondly, Look in the Tables, showing the value of Leaseholds or Annuities upon a Certainty, for the nearest equivalent Sum to the Value at each Age.

Thirdly, These Sums added together the Total will be the Number of Years the two Lives are valued at together, upon a Certainty.

Lastly, Look for the Number of Years both Lives are valued at together, in the Tables for valuing of Leasehold Estates or Annuities for a certain Term of Years, and it will show how many Years, Months, etc., Purchase an Annuity upon two Lives is worth.

[4] Fernand Faure, *The Development and Progress of Statistics in France* (The History of Statistics collected and edited by John Koren for the American Statistical Association, New York, 1918, pp. 244 *sq.*).

in 1599) made great efforts to get detailed information
as to the debts and revenues of the King, as a preparation
of his great financial reform,[1] and a Cabinet d'Archives
was established, where all the documents concerning the
administration of finances should be preserved. Unfortunately Sully's successors took very little interest in
work of this kind, and it was not till Colbert became
Contrôleur général (1665) that similar efforts could be
taken up. Colbert required reports from the *intendants*,
not only on taxes (particularly *la taille*), but also on commerce, manufacture, and principally on population.
Thus the district of *Tours* having had a heavy mortality
in 1663, the *intendant* is asked for the "number of inhabitants compared with the number three or four years
ago"; a letter to the *intendant* of *Alençon* inquires about
the causes of the increase and the decrease of population.[2] From 1670 the number of *actes de l'état civil*
(baptisms, births and burials) for the City of Paris was
published yearly. In fact, as early as in the beginning
of the fourteenth century marriages and deaths were
regularly registered in certain regions of *Burgundy*, and
in 1579 the curates of all the parishes of France by
the Ordinance of Blois were ordered to keep registers
of baptisms, marriages and burials. This is the foundation-stone of our knowledge as to the population of
France in former times, even though the original object
with regard to this registration was principally of fiscal
or judicial nature, a statistical treatment being at that
time not within the horizon. Knowing, for instance,
the total number of baptisms in a whole country—a
number which again came very near to the number of
live-births—and having, moreover, the birth-rate in certain districts at their disposal, it was possible for the
political arithmeticians to calculate the population with
tolerable accuracy. In this respect France seems to
have had better conditions than England.

---

[1] Much material of this kind is to be found in Forbonnais, *Recherches et
Considérations sur les Finances de France depuis 1595 jusqu'à 1721*, Liége, 1758.

[2] Faure, l.c., p. 248.

There was another source of information which naturally presented itself, viz. the number of households or hearths (*feux*) in the various districts of France. Looking through the reports from the *intendants* from 1683 to 1715, we find much material which it unfortunately is impossible [1] to treat from a statistical point of view, though it throws light on the social and economic history of France under Louis XIV. Many letters from the *intendants* testify to the great misery in the realm, but mostly in general and vague terms; they complain of the bad condition of agriculture, or they make suggestions as to a reduction of taxes, etc. Sometimes, however, positive statistical information is required from the *intendant*, and then we are struck by the remarkable lack of uniformity. Fernand Faure (l.c., pp. 254 *sq.*) describes the material in the *mémoires* from 1698. The *intendants* had been asked to give data as to the population, but they followed quite different principles. Some of them tried a direct enumeration of the whole population, head by head, or with exclusion of certain classes. Others counted the households, several of them taking their figures from the tax registers, particularly from the "Capitation," 1695, and here again sometimes only the taxable hearths were enumerated, whereas in other cases the whole number was given.

No wonder that the authors of this period were unable to overcome these difficulties. We may here refer to a voluminous work published by Saugrain (Imprimeur-Libraire, Juré de l'Université de Paris).[2] He gives details for each parish in the realm, as a rule with the number of *feux*, but without any system, the fiscal view evidently having been predominant in the reports from which the details have been compiled. For Paris, which he calls the most considerable place in the world,[3] the number of inhabitants is given, viz. 750,000, including

[1] *Correspondance des Contrôleurs Généraux des Finances avec les Intendants des Provinces*, publiée par A. M. de Boislisle, 1874–98.

[2] *Nouveau Dénombrement du Royaume par Généralités, Elections, Paroisses et Feux*, Paris, 1720.

[3] Ville Capitale du Royaume et la plus considérable de l'Univers.

150,000 *domestiques*. Probably this is an overstatement, but here no particulars are given as to hearths. We shall also look in vain for a system in Boulainvilliers' great work with abstracts of the *mémoires* of the *intendants* from the close of the seventeenth century.[1] The first edition appeared fifteen years after the death of the author. It contains many details as to hearths, occasionally also as to the numbers of inhabitants, for instance in Alsace, but there is no attempt at a general survey.

It was of course a drawback for the progress of statistics that the authors had frequently no permission from the absolute Government to publish their works, if indeed they had any access to the archives where the material was hidden. It was perhaps easy enough to get the permission if the work was sufficiently neutral, or—still more so—if it contributed to the glorification of the monarchy. This was the case with a work published by the Benedictines in 1749.[2] It contains many details as to civil and military officers, peers, knights of various orders, ceremonial at the court, etc., and it may be valuable to students of the administration under *l'ancien régime*, though hardly of interest to a statistician. But if a book earnestly criticised the system and suggested remedies against the social evils of the time, the obstacles were very great. The fate of Vauban's *Dixme royale* was mentioned above. Sometimes the works circulated in manuscript, sometimes they were secretly printed, or they found an asylum abroad. Forbonnais' above-quoted work was published in Liége and Basle.[3] Marquis d'Argenson's work on the Government in France, which however does not belong to statistical

---

[1] *État de la France dans lequel ou voit tout ce qui regarde le Gouvernement Ecclésiastique, le Militaire, la Justice, les Finances, le Commerce, les Manufactures, le nombre des Habitans, et en général tout ce qui peut faire connaître à fond cette Monarchie*, London, 1737, 2nd ed., 1752.

[2] *L'État de la France.* It was dedicated to the King by "Les très humbles, très obéissans et très fidèles Sujets, les *Religieux* Bénédictins de la Congrégation de Saint-Maur."

[3] In Basle it was printed anonymously after a manuscript copy, the publisher saying in his preface that he had been unable to find the author.

literature, was published in Amsterdam[1] seven years after his death, after several years' circulation of manuscript copies. And Boulainvilliers' above-quoted *État de la France* found a publisher in London.

The problem of finding the population for the whole realm quite naturally met with great difficulties on account of the suspicion of the common people. There were optimists who regarded this problem as very easy. This was the case with Fénélon,[2] in whose opinion it was just as easy for the King to know the number of his people as for the shepherd to know the number of his flock, "He has only to wish to know," and we have seen that Vauban and Leibnitz shared his opinion. But in practice it was otherwise. Quite naturally the population looked upon a census as the preparation for new taxes. As late as in 1789 the Chevalier des Pommelles says that the people have so many prejudices against an enumeration of the inhabitants that in 1786 the Provincial Assembly of Auch was obliged to stop it in the province on account of the disturbance it caused.[3]

[1] *Considérations sur le Gouvernement ancien et présent de la France*, Amsterdam, 1764.

[2] Fernand Faure, l.c., p. 252.

[3] *Tables de la Population de toutes les Provinces de France . . .*, Paris, 1789, p. 45.

CHAPTER VI

PROGRESS IN THE MIDDLE OF THE EIGHTEENTH CENTURY

22. AFTER the stagnation in the first part of the eighteenth century a new era opened in the history of statistics. *Sweden* deserves first mention, as this country was the first to give political arithmetic a solid basis by a system of official statistics. Direct observations were also made in *France* and *Holland* (annuitants, members of tontines, monks and nuns). And though the Bills of Mortality in *England* made little progress, the observations at hand were at least treated more critically than before, as shown by Th. Simpson's mortality-table for London.

In *Sweden*, as in most other countries, it was felt as a great drawback that the population was too sparse to undertake all the economic tasks arising within the kingdom in agriculture and industry, not to speak of the problem of raising the military forces judged necessary for the protection of the country. Anders Berch's remarks on the population problem were mentioned above (Art. 19). In 1744 Salander wrote on the same problem.[1] If France had 20 millions, Sweden had probably only 3 millions. But he maintains that the soil could produce sufficient food for 20 millions. If this appears doubtful, he is willing to reduce the number to 10, and even to 5 millions, but at all events he had no doubt whatever that agriculture in Sweden could produce sufficient food for these 5 millions.

It is evident that loose estimates of this kind could

[1] Försök, om Sverige kan ved egen Växt föda sina Invånere" (*Kongl. Wetenskaps Academiens Handlingar*, Vol. V, 1744).

not in the long run prove satisfactory to the discussion
of the population problem, and after long deliberations
a bill for making tabular records of the population was
approved by the King and became law on February 3rd,
1748. For many years the Swedish clergy had had
the duty of keeping parish registers, containing lists
of the members of the congregation, of marriages,
births and deaths, and of persons entering or leaving the
parish. Incomplete as these lists may have been, they
quite naturally became the foundation-stone of the
Swedish system of official vital statistics. Occasionally
they were used for local statistical investigations, as in
a paper by Wassenius on the vital statistics of a parish.[1]
But a deeper investigation was made by the mathematician Pehr Elvius (1710–49), Secretary of the
Swedish Academy of Science, who had undertaken to
compile lists of births and deaths for the whole kingdom
in order to determine the probable number of inhabitants. The result was a report (1746) which the
Academy sent to Parliament and which probably played
a significant part in prevailing on the Parliament to
pass the above-named Act.[2] As Elvius, like most
political-arithmeticians of those days, treated his observations freely, it is not possible to reconstruct his calculations with absolute accuracy, though his method is on
the whole clear enough. He regards 70,000 as the
normal yearly number of deaths, and he distributes

---

[1] "Wassenda Församlings Forökelse genom födda och vigda, så ock aftagande
genom döda . . ." (*Kgl. Wet. Acad.*, Vol. VIII, for 1747).

[2] As to the History of Swedish official statistics, see E. Arosenius, *Bidrag
till det Svenska Tabelverkets Historia*, Stockholm, 1928, and the same author's
The History and Organization of Swedish Official Statistics (The History of
Statistics collected and edited by John Koren, New York, 1918).

A. Hjelt, *De första officiela relationerna om svenska tabellverket, åren* 1749–57,
Helsingfors, 1899.

A. Hjelt, *Det svenska tabellverkets uppkomst, organisation och tidigare verk-
samhet*, Helsingfors, 1900.

Lundell, *Den politiska arithmetikens uppkomst och utveckling*, Helsingfors,
1911.

G. Sundbärg, *Bevölkerungsstatistik Schwedens*, 1750–1900, Stockholm, 1907.

Further, *The Official Vital Statistics of the Scandinavian Countries and the
Baltic Republics* (League of Nations, Health Organisation, Geneva, 1926).

them according to age after observations made in certain parts of the kingdom. Following in the main Halley's method for finding the number of inhabitants in Breslau, on the supposition that the population was stationary, he would, for instance, proceed thus: Out of the 70,000 persons who died yearly, about one-third were under 3 years, and about 29,300 under 10 years of age. Supposing, in order to arrive at the same numerical result as Elvius, that 8,000 died between 3 and 10 years of age, and 21,300 under 3 years, we shall find that 70,000 newly-born altogether have passed about 178,000 years before 3 years of age, viz. 1·5 (70,000 + 48,700), and about 313,000 between 3 and 10, viz. 3·5 (48,700 + 40,700). He thus calculates 491,000 years of life, which, on the hypothesis of the stationary population, is the same as the number of persons living under 10 years of age. For the following ages he uses decennial averages. The total number of inhabitants of all ages he finds to be 2,097,000, which is not far from the number which some years later was found by direct observation.

Elvius did not regard this result as more than a pre-liminary attempt, and was himself an advocate of a regular system of official vital statistics, such as was established in 1748. According to this system, some-what complicated schedules had to be filled out every year for each parish. Thus there were particulars, for each calendar month, of *baptisms* of legitimate and illegitimate children, with distinction as to sex, further, *weddings* and number of marriages dissolved by death, the number of *deaths*, separately for each sex, for children under 10 years of age, unmarried and married people, with additional remarks on still-births, plural births, etc. Another table gave a classification of *deaths according to age, sex and cause of death*. The children were classified according to age between 0 and 1 year, 1 and 3 years and 3 and 5 years respectively, then follow 17 quinquennial age-groups and, finally, deaths above 90. This detailed combination of age and cause of death was indeed a remarkable step forward in vital statistics.[1] The nomenclature of the causes of death, 33 in number, would not of course satisfy a modern medical statistician, but it gives much interesting information about that period, notably with regard to violent deaths and zymotic diseases. The first report, for 1749, shows 12 per cent. of the deaths as due to small-pox and measles, 6 per cent. as due to scarlet fever, and 5 per cent. to whooping-cough (all of which causes, from 1911 to 1920 inclusive, only counted 1 to 3 per cent. of the total deaths); 14 per cent. died from consumption and lung disease. Several secondary causes of deaths were noted, such as dropsy and jaundice.

Finally, there were details with regard to the *population*. The number of persons of each sex in the same age-groups as the deaths was required. Further, the population was distributed as to sex and conjugal condition (children under 15 years, single, married, widowers and widows), and according to occupation and rank (sometimes with particulars as to children under 15 years, and young persons). In addition, the number of households, of inns and public-houses, etc., was given. It is interesting to note the average number of persons belonging to each household, which, compared with modern times, was high (six to seven members against three to four nowadays).

To a certain extent the magistrates in the towns took part in preparing the lists of the population, but the main work devolved upon the clergy, the study of a pastor being in fact a small statistical bureau. The clergy complained bitterly and obtained relief to the extent that the lists of population were only required to be prepared every three years. It could not of course be expected that the material which was sent by the clergy should be faultless, in fact there were complaints that the lists were often very imperfect. Perhaps not the least trouble, especially in the towns, was caused

---

[1] Sundbärg has calculated rates of mortality according to age and sex, from *tuberculosis*, 1776–1800, l.c., p. 148, Table 60.

by the particulars which had to be entered on removal to and from a parish.

A serious drawback was the decentralisation which characterised the whole institution. From the parish the lists were sent to the deaneries, where they were summarised. These summary tables were again forwarded to the consistories which had the duty of sending their condensed reports to the provincial governors for that part of the diocese which belonged to his province. Finally, the provincial government sent a summary for the province to the "Kanslikollegium."[1]

23. Pehr Elvius died in 1749. "Kanslisecretary" E. Carleson was given the task of preparing a general summary, with the assistance of some of the members of the Academy of Science. The most active of these was the astronomer Per Wargentin (1717–83), the successor of Elvius as the secretary of the academy. By a royal rescript of 1756 this committee was made a permanent institution, the Tabular Commission ("Tabell-kommissionen").

Before the latter event Wargentin had published some papers in which he treated the death-lists in 1749.[2]

So far there is nothing new in these papers as Wargentin only uses well-known methods. But it was very important that here for the first time detailed observations embracing a whole country were at hand, and, moreover, Wargentin gives the programme of future investigations. He maintains that it would be possible to draw safe conclusions, if not only the numbers of births, marriages and deaths were known but also the numbers of the living. He adds that Halley has shown a short-cut, and the following investigations are based on his method. He does, not, however, describe it very clearly, so that it may be doubtful whether at

---

[1] Arosenius, *The History and Organization of Swedish Official Statistics,* l.c., p. 541.

[2] "Om nyttan af Förteckningar på födda och döda" (*Kgl. Vet. Acad.,* Vol. XV, for 1754, and Vol. XVI, for 1755).

that time he had thoroughly understood it, but in a later investigation (1767) he gives a quite clear statement.

Wargentin gives the distribution per mille of deaths 1749 according to age (still-born included) in the whole Kingdom and in six provinces separately ("Höfdinge dömen"), where the frequency of epidemic diseases had been relatively small. But of these two tables combined with certain other observations (Halley, Kersseboom, Deparcieux) he forms a new table. This is interesting inasmuch as it explains how Süssmilch some years later calculated the mortality-table, which carries his name and which in spite of its evident defects enjoyed a high reputation. It will be mentioned below.

Wargentin looks upon Halley's table as trustworthy even for Sweden, as regards mortality, and by comparing the figures he draws various conclusions with regard to the growth of the population in Sweden. His remarks are, however, not very clear.

The Tabular Commission took up its task with great zeal; two reports from 1751 and 1761 gave various results of the work, with more general remarks on the population problem, particularly burning in Sweden which, as may be surmised, had suffered severely from the long wars at the beginning of the century. The first report complained very strongly of the many difficulties which must be felt in so thinly populated a country. Many persons died yearly whose lives might have been saved; it was urged that there ought to be public access to medicinal drugs, that inoculation against small-pox ought to be tried, that too many children died in infancy through overlaying, that there were too few physicians. The report also suggests that students of divinity should acquire some knowledge of medicine before being appointed as clergymen. Further, that there ought to be more inducement to marry. It was a pity that so many thousands were forced to live as beggars (according to the lists there were 29,000 "paupers not in hospital"). The second report considered in detail the problem of emigration. In the course of three years 24,000 persons had emigrated; the report complained of the emigration to Copenhagen, to Norway and to Russia. Later, however, Wargentin acknowledged that the supposedly high emigration-figures were largely due to inaccuracy of the registers.

The report also drew certain conclusions with regard to the causes of *fertility*. Opulence and luxury were a hindrance to fertility, the number of conceptions being relatively small in the months when the peasant had the most plentiful supply of provisions.

These reports were not published.[1] As in so many other countries statistical information, particularly on population, was regarded as a State secret. In Sweden, particularly, it was feared to allow the serious complaints as to deficiency of population to be known in the neighbouring States. Gradually, however, this silence was abandoned, and several statistical observations were printed in the proceedings of the Academy of Science, so intimately connected with the Commission. Various papers were quoted above. In 1762 the Commission finally got permission to publish a yearly report in the Acts of the Academy, and various important results were published in the following years; in 1764–5 the Secretary of the Commission, E. F. Runeberg, gave the number of inhabitants in Sweden and Finland in 1760 (2,360,000), and in 1766 Wargentin published his famous mortality-tables for the nine years 1755–63.[2] Here, for the first time, mortality-tables for a whole country are to be found, based on the observations of the living population as well as on deaths. Wargentin compared, for instance, the average number of deaths in the triennium, 1755–7, in one of the twenty-one classes of age, with the corresponding number of inhabitants registered in 1757, asking among how many persons one death would occur in each year. Having these enumerations at his

---

[1] They are reproduced by Hjelt in his above-quoted treatise of 1899.

[2] "Mortalitäten i Sverige i Anledning af Tabell-Verket" (*Kgl. Vet. Acad.*, Vol. XXVII). Republished by "Lifförsäkrings-Aktiebolaget Thule," Stockholm, 1930.

disposal, he might have used an interpolation in order to find the average population in the triennium, but the population was growing so slowly, that a correction of this kind would have been insignificant, and as a first attempt at finding the mortality of a whole population by direct observation, these calculations may justly be looked upon as sufficiently reliable. It must not be forgotten, that, as mentioned above, there were considerable inaccuracies which no interpolation could do away with. In his later years Wargentin made an investigation on this subject and found that whereas some of the enumerations were tolerably exact, as, for instance, those of 1751 and 1763, others showed obvious defects.

Wargentin did not go further in his investigation concerning mortality; he did not, for instance, calculate the expectation of life or the numbers surviving of a certain generation at various ages. On the whole he did not enter much into theoretical questions. But at all events he deserves praise for his contributions to vital statistics. As one important result we can refer to the fact that for the first time he proved that in a general population the rate of mortality of the female sex was smaller than that of males.

It may be added that *Iceland* already had a general census in the year 1703,[1] which was carried out on the proposal of an Icelandic Commission, appointed to inquire into the economic condition of the country. The name, occupation and age of each individual was recorded; paupers were registered separately, as also persons who were temporarily present in the place concerned but had their home elsewhere. Unfortunately there was at that time no organisation to work out the results. The census is going to be published, and it will form a most interesting basis for the study of the structure of the population in this remote country, two centuries ago.

Other Scandinavian countries also had occasional enumerations. Thus in *Norway* as early as in 1662 a

---

[1] Th. Thorsteinsson, "Den islandske Statistiks Omfang og Vilkaar," *Nordisk Statistisk Tidskrift*, 1922.

census of males above 12 years of age had been taken, for military purposes; in Denmark partial enumerations were taken in 1645 and 1660 for the purposes of taxation. But the results of enumerations of this type remained mostly unpublished and they could therefore have no influence on the evolution of political arithmetic. This was also the case with several local or general numerations in various countries outside of Scandinavia in the course of the seventeenth and eighteenth centuries.

24. In the same year in which Pehr Elvius published his investigation a highly interesting work on mortality appeared in *France*, by Deparcieux (1703–68).[1]   Here old and new methods come in use.   Thus in order to calculate mortality-tables for monks and nuns he distributes the deaths between 1685 and 1745 in several cloisters, according to age, under the supposition that the number of inmates in these institutions were on the whole constant.   But for the Benedictine monks separately he calculates a table for the period 1607–1745 on a thoroughly correct principle.   The monks concerned entered in 1607–69, aged 17 to 25, and all of them died before 1745.   As none of them left during the period of observation it was easy to calculate the numbers of persons exposed to death at each age, and consequently to find the proper rates of mortality.   The next step would be to take the fluctuations of the numbers into consideration in cases where members are leaving during the period of observation, as also to find the numbers of years, where several members are still alive at the close of the period.   This would have been necessary if the period 1607–1745 had been divided into various periods in order to observe changes in mortality, a problem which, however, quite naturally did not occur to political arithmeticians in this epoch.   But Deparcieux also mastered this last step: he calculates mortality-tables for the members of two tontines, of 1689 and 1696, embracing 5,911 and 3,345 persons respectively, of whom, in 1745, 711 and 616 respectively were still alive.

---

[1] *Essai sur les probabilités de la durée de la vie humaine*, Paris, 1746.

This is done quite correctly, the numbers exposed to risk being calculated from the third year of age.

In order to calculate the mean duration of life, Deparcieux recommends an approximate method. In a stationary population with a regular balance of births and deaths the expectation of life can be formed by dividing the number of inhabitants by the number of births. But if there is a surplus of births he takes it for granted that the expectation of life is higher, whereas the result would be too high, if he used the number of deaths as denominator. He therefore suggests taking the mean of births and deaths as denominator. This method or similar formulas were frequently recommended in the following 100 years. The calculation cannot of course give very exact results. Thus we find that in Denmark 1840–9 Deparcieux's method would give 39 years compared to 42·2 years correctly calculated. 1921–5 would give 60 and 61·1 years respectively.

After having published his chief work Deparcieux did not contribute much to political arithmetic. He had a discussion with an author (Thomas) who had criticised his results, and in 1760 he published a supplement to his book, with various observations, for instance, Swedish death-lists 1754–6. It appears from some recently published letters to Wargentin (1760–7)[1] that he was planning a second revised edition. This is also mentioned by a nephew of his who in 1781 published a treatise on annuities.[2] With the assistance of the bishops Deparcieux had obtained death-lists from 162 priests in various parts of France for sixteen years. Evidently he wished in this way to get a broader foundation for mortality statistics, thus meeting objections concerning the applicability of his observations from the narrow sphere of tontines and religious institutions, even though he did so at the expense of correctness of method. But death prevented him from realising his plan, and attempts

---

[1] Arosenius, *Bidrag till det Svenska Tabelverkets Historia*, 1928, pp. *38–*45.

[2] Deparcieux, *Traité des annuités, accompagné de plusieurs tables très utiles*, Paris, 1781.

made after his death at treating the material received from the priests proved unsuccessful.

Other important contributions to political arithmetic were published in *Holland*. Thus Struyck (1687–1769) published an interesting work, chiefly containing contributions to astronomy and geography, but with several observations on vital statistics.[1] Besides detailed observations he naturally tries to give various estimates. The world population is supposed to be 500 millions, and he estimates the number of deaths in every hour to be 2,000 (35 per mille yearly). But as to other questions his basis is more solid. He discusses ably the difference of mortality of males and females, partly based on German observations. This he further elucidates by observations on annuitants, following these quite correctly, according to age, from quinquennium to quinquennium, thus finding the number exposed to risk for each age and sex.

In a later publication[2] Struyck collected several observations on vital statistics in Holland and other countries. There are several other interesting observations, on mortality in childbed with distinctions as to time passed between the birth and the death, further on twins and multiple births, on mortality of sailors on the long journey from Holland to the Cape (which was extraordinary in comparison with modern observations). He is interested in the problem of climacteric years, and here as in other directions he shows much critical sense.

In statistical literature, Kersseboom (1691–1771) is perhaps more frequently quoted than Struyck, although his contributions to vital statistics appeared in a less regular form, being partly of a polemic nature. These contributions were published 1737–48.[3]

---

[1] *Inleiding tot de Algemeene Geographie benevens eenige Sterrekundige en andere Verhandelingen*, Amsterdam, 1740.

[2] *Vervolg van de Beschryving der Staartsterren, en nader Ontdekkingen omtrent den Staat van't Menschelyk Geslagt*, Amsterdam, 1753.

[3] See, for instance, three *Verhandelingen over de probable Meenigte des Volks in de Provintie van Hollandt en Westfrieslandt* (1738–42) (later republished under the title : *Proeven van politique rekenkunde*, 1748). A good account of his contributions to political arithmetic was given by G. F. Knapp, *Theorie des Bevölkerungs-Wechsels*, Braunschweig, 1874.

Kersseboom has not the mathematical training of which Struyck was possessed, though his conclusions are generally clear. Owing to the form of his publications a complete system cannot be expected, he was too much engaged in attack and defence, in criticism and anticriticism. It may even occur that he uses in one paper a method of which he thoroughly disapproves in another one, as, for instance, when he criticises van der Burch for having calculated a mortality-table from the London Bills, though the number of inhabitants could not be found in that way, and two years later uses the very same observations to calculate a mortality-table and the population.[1]

Van der Burch (1673–1758) did not in fact deserve this bitter criticism; he seems to have written clearly on the problems he took up, as when he treated annuities from a financial point of view, maintaining that the price of an annuity ought to depend on the age of the annuitant.[2]

Even though this desultory production prevented Kersseboom from a systematic treatment of the problems concerned, his thoughts are always circling around the same ideas; he is constantly returning to the question how to find the population from the number of births and the mean duration of life. He is particularly interested in estimating the number of inhabitants in the provinces of Holland and Westfriesland. In order to reach this goal he examines a rather large amount of material available concerning annuitants. Like Struyck, he finds quite correctly the number of persons who have been exposed to risk at each age, thus arriving at a life-table for well-to-do people in his country in the seventeenth and the earlier part of the eighteenth century. He is not to be blamed for having generalised these results, so that he considered that his table would hold good for a whole population, and not for annuitants

---

[1] *Observatien waarin voornamentlyk getoont word wat is gelijktijdigheid,* 1740; *Derde Verhandeling,* etc., 1742.
[2] *Bouwstoffen,* pp. 155 *sq.*

only. At all events, his mortality-table for annuitants deserves the reputation it enjoyed, at least for all ages above one year. As to infant mortality, he calculated on numbers for the whole population, but these were only directly observed in some cases and depended partly on estimates, though he took much trouble to prove the reliability of his figures. The supposed annual number of births in the two provinces was 28,000, out of which 5,500 were supposed to die in the first year of life. His life-table begins with 1,400, or a twentieth of the annual number of births given, and from this table he finds the expectation of life to be 35. He seems to have found this number by adding up the numbers of living at the beginning of each year of age, without taking the decrease in the course of the year into consideration.

Having found the mean duration of life, he calculates the birth-rate at 29 per mille, 35 being the number of inhabitants with one birth yearly. 29 per mille is rather a low birth-rate for that period, and Kersseboom did not escape criticism on account of this point. Probably the rates of mortality in the general population were higher than among annuitants, and if he had had a suitable mortality-table at his disposal he would have found an expectation of life lower than 35, even if he were right in supposing the population to be stationary.

As usual with political arithmeticians, he seems to have treated the observations freely, it being therefore impossible to reconstruct quite exactly his calculation as to the age distribution of the population. As regards the first five years of age, the number alive according to his supposition seems to be somewhat too high. Comparing his age distribution with the Swedish numbers in the middle of the eighteenth century, we find a pretty good correspondence. As we should quite naturally expect, the numbers above 80 seem to be relatively high judged by Kersseboom's table.

As with other political arithmeticians of his time, he is inclined to transfer results from one country to another. Thus he uses King's estimates of the distribution of the population according to matrimonial classes as reliable also for Holland.

Sometimes his conclusions are a little clumsy. In one of his papers [1] he deals with observations (for each sex separately) on annuitants who entered at the same time and at the same age, and of whom none was alive at the time when the investigation was made. Summing up the ages which all the annuitants had reached, he found the mean duration of life. But he failed to use his material to calculate the rates of mortality.

In spite of all objections, however, Kersseboom justly deserves his reputation as one of the most prominent statisticians of the eighteenth century.

25. While great progress was made in this epoch on the Continent with regard to statistical observations, political arithmetic in *England* was chiefly engaged in revising and improving the material at hand. Though annuities on life were well known, no attempt was made at treating the material of the life-offices, nor did Parliament follow the example of Sweden regarding a system of official statistics. An important reform was made with regard to the Bills of Mortality, in so far as the age distribution of the deaths was known from the year 1728. The material was, however, rather defective, many deaths—particularly of dissenters—were not registered. Interesting comments on these defects were made in an anonymous introduction to a work which was published in London, 1759: *A Collection of the Yearly Bills of Mortality from 1657 to 1758 inclusive.* In a treatise included in this work Corbyn Morris recommends various important improvements.[2] He proposes, for instance, to divide the deaths of infants into three groups, viz. under 1 month, 1–3 months and 3–12 months. Further, the lists ought to give the distribution according to *birth-year* as well as to *age*. He complains

---

[1] *Derde Verhandeling . . .,* 1742.

[2] *Observations on the past Growth and present State of the City of London,* 1751, 2nd ed., 1757.

very much of the disproportion between the number of christenings registered and of births.

Even if improvements of this kind could be obtained the observations had the obvious defect that *migrations* had an influence on the age distribution of the deaths which made it impossible to apply Halley's method directly, especially in a place like London, where immigration was so conspicuous. First of all, the mathematician Th. Simpson tried to find a solution in a work on annuities.[1] He tries to correct the observations by combining Smart's table for London, 1728–37, with Halley's for Breslau. He considers the immigration after the age of 25 as insignificant; above this age therefore Smart's table can be used without amendment, but for the younger years a correction in respect of this afflux from outside will be necessary. He does not explain his method clearly, the main result being that 140 per mille of all the deaths in London occur among immigrants. He asserts that he has compared the number of christenings and burials and observed by help of Halley's table "the proportion which there is between the degrees of mortality at *London* and *Breslaw* in the other parts of life, where the ages are greater than 25."

Simpson seems to have simplified the problem by supposing that immigration into London was concentrated around the age of 25. If 140 out of 1,000 deaths belonged to immigrants and 860 to natives, the deaths could be divided into two sections, one pertaining to the age under 25, altogether 574, natives solely, and the other to 426 above 25, out of whom 286 were natives and 140 immigrants.

In order to combine these two sections into one correct table of mortality he could either reduce all the numbers above 25 in the proportion of 286 to 426, or he could raise all numbers below that age in the proportion 426 to 286. Choosing the latter alternative, he will have to increase the 860 native new-born by 49 per cent., the table thus beginning with about 1,280.

---

[1] *The Doctrine of Annuities and Reversions*, London, 1742.

A similar method was used several years after by Richard Price in his Observations on Reversionary Payments.[1] He wished to calculate a life-table for London on the basis of the numbers of deaths 1759–68, and supposed that about half of the persons dying in London above 18–20 were immigrants who settled at that age. On similar principles tables were constructed for *Northampton* and *Norwich*. These two tables were much appreciated, although the material was rather limited. In Norwich the distribution according to age below 10 was unknown. Price therefore uses a proportion similar to that in Northampton.

While Simpson and Price tried to take *migrations* into consideration, another attempt at improving the death lists was made on the Continent by the famous mathematician, Euler (1707–83), under the supposition that migrations were trifling, whereas there was a regular *surplus of births* over deaths.[2] If, for instance, the yearly surplus is 1 per cent. of the population, and if the rates of mortality can be considered as constant, then persons aged 70 will belong to a generation about half as numerous as the generation born at the present moment. It will thus be easy to find exact rates of mortality from the numbers of deaths, if the yearly rate of surplus is known.

These are the first attempts at taking the movement of the population into consideration by the construction of mortality-tables. Generally, political arithmeticians took it more easy. This is the case with the French naturalist, Buffon (1707–88), who in his *Histoire Naturelle* (*De l'homme*, Tome II, 1749) deals with a material which Dupré de Saint-Maur had placed at his disposal, containing lists of deaths in some parishes in Paris and outside of it. He takes no notice of the influence of

---

[1] First edition, 1771, here quoted after the third edition, London, 1773. Confer Sutton, "On the Method used by Dr. Price in the Construction of the Northampton Mortality-Table" (*Journal Institute of Actuaries*, XVIII, 1875).

[2] "Recherches générales sur la mortalité et la multiplication du genre humain," Berlin, 1767 (*Histoire de l'académie royale des sciences et belles-lettres, année* 1760).

the movements of the population. He is chiefly interested in the probable duration of life—the age at which just half of the generation in question will be alive ("les probabilités de la durée de la vie"). In other respects his treatise is not uninteresting, with useful remarks on the concentration of the numbers on the round years of age, and his observations on the numbers of foundlings throw light on the social history of that epoch.

Other authors of this period may be read with interest. In Holland N. Duyn, who died 1745, wrote upon the influence of seasons on mortality.[1] Th. Short, in a work from 1750,[2] takes up a number of problems, though unfortunately in a rather superficial way. Thus he tries to show the influence of the geologic situation and of the soil on mortality. If circumstances are favourable in this respect, the number of miscarriages will be relatively small, and there will be many male births. Astronomic phenomena have influence on mortality, particularly eclipses. Children born in the cold months have a greater mortality than others. Like Wargentin, he maintains that "the Months of the greatest Ease, Repletion, Insolence, and the smallest Discharge, are most improper for Procreation" (l.c., p. 143). The months of October, November and December, with relatively few births, have proportionally many male births, "for strong Labour and Exercise has strung the Nerves and purified the Blood."

---

[1] *Bouwstoffen,* pp. 167 *sq.*

[2] *New Observations, Natural, Moral, Civil, Political and Medical, on City, Town, and Country Bills of Mortality . . . with an Appendix on the Weather and Meteors,* London, 1750.

CHAPTER VII

SUSSMILCH AND HIS CONTEMPORARIES

26. MANY of the conclusions drawn by the political arithmeticians of those days were ill-founded and premature and their methods often very imperfect. Still, to some extent these results justified the high opinion with regard to the new science which frequently found expression in literature. It will therefore be worth while to get a bird's-eye view of what the statisticians of that epoch knew or believed they knew.

First we must mention the *regularity* of certain phenomena in vital statistics, which deeply impressed political arithmeticians of those days. These phenomena were frequently looked upon as proofs of a Divine Order. This was particularly the case with the question of the balance between the two sexes. Thus Thomas Short in his above-mentioned *New Observations* finds a remarkable providence in the small surplus of male births; as greater dangers menace boys than girls, a balance between the sexes would be the result, and he is led to the conclusion that "Polygamy is a most ridiculous, monstrous Custom." Wargentin writes in the same vein.

Two clergymen who contributed to political arithmetic maintained with particular force that the regularity is the result of Divine Providence: Will Derham (1657–1735) in England, and J. P. Süssmilch (1707–67) in Germany. The former published a frequently quoted series of sermons under the title: *Physico-Theology.*[1] He

---

[1] *Physico-Theology or a Demonstration of Being and Attributes of God from the Works of Creation,* 1713. Here quoted after the eighth edition, London, 1732. Cf. James Bonar, *Theories of Population from Raleigh to Arthur Young,* London, 1931, pp. 136 *sq.*

did not, however, collect many statistics, as did Süssmilch, who, in 1741, published a work on the *Divine Order* which appeared in a very enlarged edition in 1761.[1]

Süssmilch collected material with great care wherever it was accessible. His correspondence with Wargentin[2] testifies to his energy, and his work is a fairly complete compendium of all the statistical literature up to his time.

He is not a very original author. Having no mathematical training, he was naturally led to accept without much criticism the results which other authors, such as Deparcieux, had found. He is, however, by no means lacking in critical judgment, and many of his remarks show a sound common sense.

Süssmilch belongs to the army of authors in the eighteenth century who considered the population question the most important object of politics. His work contains in fact a discussion of all the causes which might influence the growth of population. His starting-point is chosen from the Bible (Gen. i. 28): "And God blessed them, and God said unto them: Be fruitful and multiply and replenish the earth, and subdue it" (Introduction, § 1). In order to reach this goal a Divine Order is established, regulating the proportion between births and deaths so as to give a normal surplus as the result. He maintains further that if there is a law of mortality, there must also be an order in diseases. It is true that epidemic diseases may cause some disturbance if certain years only are considered, but these irregularities disappear if twenty years or more are combined. The balance of sexes is a chief point in his discussion of the Divine Order: if, for instance, one male were born for every three females, then polygamy would be the natural consequence, but the wise Creator gave a proportion which led naturally to monogamy.

---

[1] *Die göttliche Ordnung in den Veränderungen des menschlichen Geschlechts aus der Geburt, dem Tode und der Fortpflanzung desselben erwiesen.* After his death the work was republished by his son-in-law Baumann, also a clergyman, who added a Supplementary Volume (Berlin, 1775-6). This fourth edition is here quoted.

[2] Reproduced by Arosenius, l.c., pp. *63-5.

In vital statistics Süssmilch finds (§ 12) a constant, general, great, perfect and beautiful order. He maintains that the duration of life is constant (§ 13). As it was 3,000 years ago, so it is in Europe nowadays: in Finland, Sweden, England, Holland and France. On account of this supposed permanence everywhere he feels justified in trying to calculate normal rates of mortality on the basis of various observations (§ 35), viz. in the country in "good years" 1 : 42 to 1 : 43, in "mixed years" 1 : 38, and as a general mean 1 : 40. Smaller towns have a mortality rate of 1 : 32, larger places, such as Berlin, 1 : 28, and still larger places, Rome and London, 1 : 24 to 1 : 25. For a whole province he considers 1 : 36 as normal, and he adds with great deference (§ 42) that every age and sex, all professions and all diseases, must contribute their share in order that the rate of mortality —one out of 36—may be the result. Dropsy, as well as convulsions and fevers, all have their part in the great tribute to the grave.

It is easy enough in our days to criticise these assertions. We find the regularity less striking than Süssmilch and his contemporaries did. In explaining his life-table he repeats his remarks on the *Divine Order*. He was struck by the harmony between the observations on monks and nuns in Paris and peasants in Brandenburg. But in fact the variations were not inconsiderable. The average for the second year of life in the country, viz. 78, was, for instance, calculated from the following four numbers: 49 and 59 respectively, 100 and 104. And as to the alleged constancy of duration of life Süssmilch had no reliable observations at all on mortality in ancient times. Modern life-tables show an enormous increase of the expectation of life, of which no eighteenth-century writer could dream.

But even though quite naturally we are induced to take notice of deviations from the average, and to ask how these irregularities can be explained, it is just as natural that our predecessors first of all were struck by the regularity and cared less for the deviations.

Süssmilch devotes a couple of chapters to marriage—and birth-statistics. He is particularly interested in the question of how long a time a doubling of the population will require. Here he was assisted by Euler, who furnished him with a table which showed the doubling period under varying circumstances. Of chief importance to Süssmilch was the probable increase of the inhabitants on the earth before the Flood. In another chapter he discusses various obstacles for increase: the plague, wars, famines, etc.

The succeeding chapters are uninteresting to the history of statistics, in so far as Süssmilch here chiefly enters into a discussion of the economies of population from the well-known mercantilistic standpoint. He brings arguments against polygamy, discusses proposals as to supporting married couples with numerous children, as to hygienic measures, luxury, etc.

After this long digression the author resumes his statistical investigations. One chapter is devoted to the population on the earth, according to various authors and to his own estimates. He considers 14,000 millions as being within the range of possibility, but the estimated population is only 1,080 millions and consequently far behind that limit. In another chapter he treats birth-statistics, particularly with regard to sex-proportion. He finds that out of 100 births 51 will be males; this holds good for Europe, and a rather limited material from Trankebar in India points in the same direction.

27. By far the most important chapter deals with *mortality statistics*. Even though his life-table is very primitively constructed, and in spite of evident defects, it enjoyed a great reputation. In his calculations he follows Wargentin closely in the above-quoted report of 1755 on the mortality in Sweden in 1749. Süssmilch, like Wargentin, evidently looks upon a mortality-table from one section of the earth's region as holding good for the whole region. Having two such tables at disposal, even though based on material of very different proportions, they felt justified in taking the mean of the

rates of mortality as an approximation to the truth, so much the more that problems of the limits of the deviations from the mean as the size of the material varies, had not yet claimed the general attention of statisticians. Wargentin takes the mean of the deaths, distributed per mille, according to Halley's and Kersseboom's tables, the Swedish experience for the whole kingdom and for the provinces where epidemic diseases had been relatively rare in 1749. A similar method is used by Süssmilch (§ 441), the only difference being that he replaces Halley's and Kersseboom's tables by two tables calculated from observations on deaths in some parishes in Brandenburg. He misunderstood Wargentin in so far as he considers the table for Sweden as only pertaining to the provinces where epidemic diseases were relatively prominent, whereas it embraced the whole kingdom, the one table thus overlapping the other. But if all the four tables can be considered as equally good observations on the normal mortality, this objection is irrelevant. According to Süssmilch's opinion, the resulting table gives a good picture of mortality in a rural population.

Süssmilch is well aware that the supposition of a constant population does not hold good (§ 463), but he looks upon his tables as approximately correct. In fact one of the Brandenburg tables is based on 1,072 deaths (stillborn included), whereas the numbers of christenings was 1,437.

Having calculated a table for the rural population, he again follows Wargentin by adding three tables for select classes, which he supposes to enjoy the same health as the peasant population. Wargentin borrowed these tables from Deparcieux, viz. one for members of tontines, one for Benedictine monks, and one for nuns, but whereas Wargentin only gives them for comparison, Süssmilch calculates one table out of all the seven tables, as applying to select classes ("Landsleute und ausgesuchte Personen"). Deparcieux's table for tontines begins with 1,000 at the age of 3, 814 of these being alive at 20. This age again is the starting-point for the tables for monks and nuns, all the tables concerned beginning with 814. Wargentin reduces the numbers in order to compare them with the other tables; this he has done somewhat freely. Each of the tables for monks and nuns begins (at the age of 20) with 523, whereas Süssmilch's table for the rural population has 521. The corresponding number for the tontines is 537. Süssmilch, however, takes no notice of these differences, and they disappear in the mean of the seven tables. This mean consequently consists of two sections, one for the first twenty years, identical with the table for rural districts, and another for the ages above 20, slightly different from that table.

Süssmilch then proceeds to calculate tables for provincial towns. The material is much restricted, the observations pertaining to two small towns in Brandenburg, altogether with a few thousand inhabitants. The average is calculated in just the same way as above.

For the city population he takes the mean of five tables (two for Berlin, in different periods, one for Paris, Vienna and Braunschweig respectively). Finally, Süssmilch joins the three tables, for the country, the towns and the cities into one (§ 461). As the original tables mostly are based on quinquennial classes of age, he undertakes an elementary adjustment so as to distribute the deaths for each year of age.

As we have seen, Süssmilch was well aware that there were considerable differences in the health conditions among the rural and the urban populations. When joining the three tables into one he might have taken the actual numbers of inhabitants into consideration. In that period the rural population was as a rule much larger than in the towns, whereas here its weight is only one-third. Here again the idea of a uniform law of mortality seems to be predominant, in spite of the evident fact that the chances of death are different.

Having constructed his mortality-table, Süssmilch calculated the numbers of the surviving at each age, and from these numbers he draws various conclusions just as Halley or Graunt did. He does not, however, make himself quite clear with regard to the ages, for strictly speaking his numbers should show how many were alive on their birthday, not how many on an average were living between two subsequent birthdays. But after the infant age this objection is of relatively small consequence. A curious estimate of the number of married couples may be mentioned here (§ 466). According to his table there are 8,794 between 20 and 40. Altogether he supposes the number of married people to be somewhat more, about 9,000. The number of deaths being 1,000, and the population according to the supposition being stationary, the 4,500 couples will give 1,000 births annually, or in other words, one marriage will on an average give a birth in $4\frac{1}{2}$ years. Another curious estimate concerns the number of inhabitants in the old Nineveh (§ 470). He interprets what is said in Jonah iv. 11: "Nineveh, that great city wherein there are more than six score thousand persons that cannot discern between their right hand and their left hand." These words, according to Süssmilch, should be taken as a statement of the number of children under 4 years. His table shows that one-ninth of the population was under that age; consequently Nineveh had 1,080,000 inhabitants.

Süssmilch calculates the probable duration of life (§§ 473 *sq.*), asking, after how many years will half of a certain generation still be alive. He speaks as if this is just the same as the expectation of life, though he explains (following Deparcieux) the proper way of finding the mean duration of life, with the addition however that the two methods will often give the same result. His own table shows that this is a mistake; thus he finds as the probable duration of life at birth 19 years, whereas the mean duration of life according to his own calculations is 29 years.

He compares his results with various tables, frequently quoting statistical phenomena which he finds remarkable, as for instance Kersseboom's comparison of the mortality of boys and girls.

In his first edition on the Divine Order Süssmilch discussed popular ideas concerning climacteric years according to which the age of 63 (7 × 9 years) had a particular risk. Instead of this he found an increase of mortality at the round years, 50, 60, etc. Later, however, he acknowledged that these irregularities might be explained as arising from errors in the lists (§ 490).

After a digression on tontines and life-annuities he resumes mortality statistics and devotes a chapter to the causes of death, chiefly with reference to English statistics. The Divine Order is also to be found here in spite of the above-mentioned irregularities caused by epidemics. He discusses (§ 528) the burning question of inoculation against small-pox. Further, he gives observations on the influence of seasons on mortality.

A succeeding chapter contains estimates of the number of inhabitants in a large number of towns and cities based on the yearly number of deaths. Finally, he devotes some paragraphs to King's and Davenant's contributions to political arithmetic.

This review can, of course, only give the main contents of Süssmilch's work. Several interesting remarks are scattered about in the text, for instance, on the source of errors arising from the fact that several persons die away from the place where their home is (§ 51), or on the two censuses in Berlin, 1747, with a week's interval, which differed but little from each other (§ 143), or the curious discussion at the royal table (§ 469) on the probable number of persons above 80: the question being whether fifty persons of this age were living in Berlin. In great haste inquiries were made, and more than 400 were found. Süssmilch himself expected 1,147, according to his life-table.

28. C. J. Baumann, the son-in-law and colleague of Süssmilch, added a supplementary volume to the fourth edition of the *Göttliche Ordnung*. In many respects we meet quite the same views. On the whole, Baumann

accepts Süssmilch's method of calculating a life-table, and the statistical observations which he has collected —chiefly on births and deaths—do not deviate in the main from the material in the original work. Nor does Baumann's long treatise on a favourite question of the time, that of funds for widows, give weighty contributions to the discussion. And we meet the same mercantilistic ideas as in Süssmilch's work; Baumann speaks (p. 187) of money as the blood in a body, an idea which we so frequently meet in the economic literature of that period.

Still, there is evidently some progress. He discerns clearly between the probable and the mean duration of life (pp. 409 *sq.*, pp. 523, 536). His critical sense is awake to observations on macrobiots, and he gives a curious example from his own experience (p. 419). He is well aware, by the calculation of life-tables, of the disturbing influence of the surplus of the births over deaths. Therefore Süssmilch's table will not tell us the number of persons among whom 1,000 will die yearly, but, on the other hand, according to his opinion, the table shows how 1,000 persons born at the same time will gradually die out.

He corrects various numerical errors in Süssmilch's life-table (pp. 361 *sq.*), and he leaves out monks and nuns as well as tontine members from the table for rural districts. And whereas Süssmilch aims at finding normal values of the rates of mortality, Baumann asserts that there is a great difference between one country and another. In consequence he does not look upon the tables for rural districts, towns and cities as essentially alike, and the average should therefore be calculated by taking the numbers of inhabitants (pp. 367 *sq.*) into consideration. In order to get a mortality-table for the Churmark Brandenburg he would, for instance, let the table for the rural districts have the weight of $\frac{2}{3}$, whereas the other two tables were to have $\frac{1}{6}$ each.

CHAPTER VIII

ESTIMATES AND ENUMERATIONS OF POPULATION

29. SEVERAL French authors in the second half of the eighteenth century took up the problem of approximate estimates of the population in France, particularly l'Abbé d'Expilly, Messance and Moheau.

D'Expilly (1719–93) was a very fertile author. He had planned a comprehensive geographical dictionary and succeeded in completing five volumes (ending with Q).[1] For several places he gives the number of souls or of hearths. In a long article on population in which he quotes various authors, such as Wargentin (with whom he corresponded),[2] he gives the results of original investigations regarding the population of France, to a great extent achieved at his own cost. In some cases there were direct enumerations at his disposal, and in Provence, as an exception, with distinction of males and females, boys and girls under 12, male and female servants, etc. In other cases he calculates the population by means of births, deaths and marriages. As a general result he found 22 millions which he distributes according to age based on the experience for Sweden.

D'Expilly quotes approvingly the work of Messance,[3] who in turn supports d'Expilly. This author was Receveur des Tailles de l'Élection de Saint-Etienne. According to him the number of inhabitants can be found

---

[1] *Dictionnaire géographique historique et politique des Gaules et de la France,* 1762–8.

[2] Arosenius, l.c., *45–7.

[3] *Recherches sur la population des Généralités d'Auvergne, de Lyon, de Rouen et de quelques provinces et villes du royaume avec des réflexions sur la valeur du bled tant en France qu'en Angleterre depuis 1674 jusqu'en 1764,* Paris, 1766.

by means of the births, the marriages or the number of households; among these methods he prefers the first. His calculations are a little curious. Thus, having on an average 1,020 births yearly and a population of 25,025, each birth will correspond to $24 \frac{1}{2} \frac{1}{40} \frac{1}{80}$ persons, viz. $24 + \frac{1}{2} + \frac{1}{40} + \frac{1}{80} = 24 \cdot 5375$. By the usual division we find $24 \cdot 5373$. He adjusts this quantity to 25, and finds that the province of Auvergne with its 24,604 births has a population of 615,100. In order to find the population in the whole kingdom he finds the number of parishes in a select population with 59,894 yearly births, or a population, with a birth-rate $\frac{1}{25}$ of nearly $1\frac{1}{2}$ millions. The 2,152 parishes have on an average 696 inhabitants each; he reduces the number to 600 and finds that 39,849 parishes have altogether about $23 \cdot 9$ millions.

Several years after Messance, Moheau published his contributions.[1] He enumerates several methods of finding the population. Thus we can base the calculations on the consumption, on tax-lists, households or houses, and on births, marriages and deaths. Out of these he chooses births as the safest starting-point. For the whole kingdom there will be one birth out of about $25\frac{1}{2}$ inhabitants, whereas there will be about 121 or 122 for each marriage. Altogether there were in five years on an average 928,918 births yearly, the population thus being 23,687,409. The number of marriages being 192,180, he finds—with less certainty—at least 23 millions, whereas a death-rate of $\frac{1}{30}$ and 793,931 deaths will give 23,817,930 inhabitants. He thus arrives at about the same result as Messance. By means of some special observations he tries to distribute the population according to age, 7 per cent. being above 60 years, 40 per cent. below 10.

He is less original concerning mortality, the mean duration of life being according to him identical w.th the mean age at death. Imperfect as his material is, he

---

[1] *Recherches et considérations sur la population de la France*, Paris, 1778. The book was probably finished in 1774.

draws comparatively correct conclusions as to the proportion between the sexes, the climacteric years (defined as the ages at which females reach puberty and cease being fertile), and his investigations as to the growth of population are not without interest. In opposition to the current opinion that population was regularly decreasing, he maintains that enumerations in certain parts of Auvergne give an increase of $\frac{2}{43}$ in 15 years, and he concludes that if this holds good for the whole country the population will double in less than 250 years (the exact number is 229 years). According to his opinion an increase of that extent would not be improbable.

Necker also was interested in the problem.[1] He chooses as birth-rate one to 25·75, as death-rate one to 29·6, and as marriage-rate one to 113⅓. In the four years 1777–80 there was a yearly average of 940,935 births, the population consequently being 24·2 millions, the marriage- and death-statistics leading to about the same result. By taking the births in the quinquennium 1776–80 he gets a somewhat higher number. Necker does not, however, enter much into the material, nor does he explain how he found the various rates, so that it is impossible to judge of his results properly. The reader naturally grows a little suspicious on seeing that births, deaths and marriages all lead to the same result.

In the above-quoted (art. 21) work by le Chevalier des Pommelles the author chooses the same rates as Necker, but he uses an average of 10 years (966,240 births). Making a correction for Paris, he arrives at 25 millions. In order to find the proportion between males and females, the distribution as to marital condition, etc., he bases himself on the rates in selected districts where it proved possible to find the number of inhabitants. That he is sometimes tempted to rather unfounded assertions is natural, as when he tries to show that marriages are most fertile in elevated places where the air is warm and dry, the reverse being the case where the air is heavy and the soil marshy and low-lying.

[1] *De l'Administration des Finances,* 1784.

30. Looking through the various attempts at finding the population of France, we get the impression that those estimates may not have been far from the truth. But it cannot be denied that the authors have possibly not been quite unbiassed, and indeed political arithmeticians were at a loss for methods to control their results. Here the famous mathematician Laplace (1749–1827) made a long step forward. In a paper of 1786 on births, marriages and deaths in Paris he proposed to find the population of France by means of the birth-rates in certain parts of the kingdom, and he makes an investigation as to the probable limits of the deviation from the real number. After the fall of *l'ancien régime* he obtained the support of the Government for a practical experiment.[1] Thirty departments in France were chosen, and in each of these a number of places, where the *maire* was sufficiently intelligent and zealous. Here the inhabitants, on September 22nd, 1802 (the republican New Year's Day), were enumerated, altogether about 2 millions, and the number of births in the three preceding years, September 22nd, 1799, to September 22nd, 1802, was counted. It was found that there was one birth yearly per 28·352845 inhabitants. If now the yearly number of births in the French Empire within its boundaries at the moment when Laplace's *Theory of Probabilities* appeared was $1\frac{1}{2}$ millions, the total population would be 42,529,267. He further finds that there will be 1,161 to one that the error will not reach half a million.

Looking at this experiment with the experience of the following three generations in mind, we have no difficulty in finding objections. Curiously enough, in the third edition (1820) the whole argument is repeated verbally, with the only exception that Laplace, according to the intermediate regulation of the frontiers of France, reduces the yearly number of births to 1 million and the population consequently to about 28 millions. But he does not recalculate the limits, and in the following he therefore uses 42 millions as the supposed number of inhabitants.

[1] Laplace, *Théorie analytique des Probabilités*, 1812, pp. 391 *sq.*

Here, evidently, the readers' doubts will set in. How could Laplace know that the normal number of births in Greater France was $1\frac{1}{2}$ millions, and in the reduced Kingdom 1 million? In fact, few years in the course of the nineteenth century show over a million births in France; as a rule, the number was below this limit. By studying this passage in Laplace's work a suspicious reader will naturally fix his attention on these two numbers: $1\cdot5$ and 1 million of births respectively.

Again, a modern statistician would wish to see the actual numbers in the various places which the investigation embraced. The limits are calculated in accordance with the results of the calculus of probabilities, but it will be necessary to know whether the various birth-rates are grouped according to the binomial law, or whether there are several centres in the country around which the numbers are grouped. This will, of course, all have influence on the limits within which the total population is to be found.

A modern statistician might also object that the number of inhabitants ought to have been taken in the middle of the triennium from which the births had come. This objection is, however, rather irrelevant, as movements of population in those days were generally much slower than later on.

But in spite of all these objections it must be acknowledged that Laplace by his solution gave a very important impulse to statistical knowledge, even though very few persons understood the range of this experiment. The modern "representative" statistics which are based on the study of selected sections of the material ("sampling"), from which the conclusions are extended to the whole material, are in reality greatly indebted to Laplace. Curiously enough, there was a general census in France in 1801, but it seems to have been anything but a success, so that Laplace's experiment was by no means superfluous. But naturally representative statistics were laid aside for a while as in the course of the nineteenth century general enumerations gradually won the confidence of the statisticians.

31. *In England*, the birthplace of political arithmetic, it seemed more difficult than in France to get reliable estimates of population. In March, 1753, a Bill was introduced in Parliament to provide for an annual counting of the people. But this Bill met with such opposition and the proposal was viewed with such alarm that the Bill was rejected by the House of Lords.[1] Curiously enough, a very interesting private experiment was at the same time made in Scotland. The Rev. Alexander Webster succeeded in getting into communication with all the ministers in Scotland and to get an account of the whole population in the year 1755. But unfortunately this report was never published.[2] In England nothing was done in this direction except local censuses here and there. And for many years it was under discussion whether the population in England was increasing or decreasing. It was the underlying supposition in Malthus's *Essay on Population* (1798) that there was an increase, but other authors maintained the reverse; for instance, Richard Price. In vain Arthur Young suggested a census.[3] He remarks that some assert that the population declines, "that we have lost a million and a half since the Revolution; and that the decrease now continues strong; others are of a direct contrary opinion." He rejects altogether the number of houses as the basis of an estimate, "for by what rate is the number of souls per house to be determined? How is the medium to be found between the palace and the cot?" He therefore proposed a regular census every five years "unless a change of national circumstances called for variations." But the public turned a deaf ear to it. Still, in the year 1800 Sir Frederick Morton Eden had to make an

---

[1] *Official Vital Statistics of England and Wales* (League of Nations Health Organisation, Geneva, 1925), p. 17.

[2] "Note on An Account of the Number of People in Scotland in the Year 1755," by Alexander Webster, one of the Ministers of Edinburgh. *Journal of the Royal Statistical Society*, LXXXV, 1922.

[3] *Proposals to the Legislature for Numbering the People*, 1771.

estimate of the population.[1]   Knowing the number of houses, he tried by sampling to find the average number of inhabitants in each house, as well as the number of births.   The result was a total of 322,000 births and a birth-rate of one out of $27\frac{3}{4}$.   This would give a total population of 9 millions; various reasons led him to expect that the number was even higher.   Price had in 1779 maintained that the population was only 5 millions.   By that time, however, the opposition against a general census was declining.   A Bill for ascertaining the population of Great Britain was under consideration by the Legislature, and in 1801 the first decennial enumeration took place, giving 9 millions as result.

In addition, the clergy of each parish were required to prepare a statement of baptisms and burials for each decennial period from 1700 to the end of the year 1780 with distinction as to sex, and of marriages, 1754–80. Thus a foundation was laid for English official statistics.

Among the authors who took part in the efforts to solve the problem of the number of inhabitants, Will Black can be mentioned.   Although he considers a census the best way to ascertain the population, he recommends as an alternative to use the number of houses as a basis, allowing 5 or $4\frac{1}{2}$ to each house.[2]

32. In other countries it was also felt desirable to find the number of inhabitants by direct enumeration.   The constitution of the *United States of America* provided for a regular census of the population with a view to determine representation in the Congress, which had to take place every tenth year, and which was taken for the first time in 1790.   But already before this time enumerations were well known, even in the Colonial period, and had been employed in order to obtain information of value

[1] *An Estimate of the Number of the Inhabitants in Great Britain and Ireland,* London, 1800.

[2] *An Arithmetical and Medical Analysis of the Diseases and Mortality of the Human Species,* London, 1789.   On the discussion in the last decennium of the eighteenth century about the population in England, see Edw. C. K. Gonner, "The Population of England in the Eighteenth Century," *Journal of the Royal Stat. Soc.,* LXXVI, 1913.

in the administration of the Colonies. Thus in 1756 a census embracing all the inhabitants of *Connecticut* was taken, in 1764 in *Massachusetts*. The articles of Confederation as originally reported in 1776 provided for a triennial enumeration of the population as a basis for apportioning the charges of war, and even though this was altered, several of the States made an enumeration.

The census of 1790 was rather summary. It returned the number of free white males above and under 16, the free white females, and, further, the slaves without distinction of sex or age, and it has been remarked that the institution of the Census seems to have been "a political incident, little regarded at the time except as a practical means of apportioning representatives and taxes." But it was at all events a remarkable step forward.[1]

Also in *Norway* and *Denmark* interesting progress can be recorded. Both countries had a census taken in 1769, and in 1787 a new census was taken in Denmark. There was no committee appointed to prepare the census reports, the work being left to private persons. The tables concerning the enumeration of 1787 were finished in the course of three years, but unfortunately no report was published; only a few results were made known to the public. Finally, in 1797 a statistical institution was established in Norway and Denmark, a "Tabelkontor," but its influence was very small, its chief duty being to collect and revise accounts concerning taxes and other public revenues. Vital statistics had only a secondary place in its activities.

In *Sweden*, the stronghold of official vital statistics, conditions were of course much more favourable. Still, no great progress can be recorded in the latter part of the eighteenth century. For several years the Tabular Commission worked with great zeal, but at last it grew tired. Wargentin died in 1783 after a protracted illness, and

---

[1] John Cummings, "Statistical work of the Federal Government of the United States," *The History of Statistics*, 1918, p. 670; see further, Charles F. Gettemy, "The Work of the Several States of the United States in the Field of Statistics," the same volume, pp. 711–12.

the commission seemed rather inactive till it was reconstructed in 1790 and the astronomer Nicander was appointed secretary.   At the close of the century a more liberal grant was allotted to the commission, so that it could more easily discharge its duties.   The work had been simplified in 1773 in so far as the lists of population were hereafter only to be sent in every five years and only for each diocese, not for the provinces.   In 1792 an important reform was made, lists for the deaneries had now to be sent direct to the commission, thus an important step towards centralisation had been taken. After 1775 the records contain particulars concerning the age of *mothers bearing children* (in five-years age groups), a most remarkable step forward.

33.   The preceding list of political arithmeticians is not complete.   In several countries we meet authors whose names deserve mention.   In Italy, for instance, Marco Lastri can be quoted.[1]   He gives several census results for Florence, also numbers of births per month, for each calendar year, each sex separately.   The schedules are, curiously enough, arranged so that the empty spaces can be filled up by the author's successors up to the year 1850. Many of the data throw light on the evolution of Florence, though it is impossible to bring complete harmony into the whole material.

Another highly interesting treatise was published by a *Swiss* clergyman, Muret,[2] on the vital statistics of the country of Vaud.   There were 100 christenings for $79\frac{1}{3}$ deaths, and the rate of mortality was one out of $45\frac{1}{9}$.   He concluded that the surplus of births will be one out of 173 yearly and the population will double in 120 years.   Then follow some curious calculations concerning the doubling period.   375 mothers have altogether borne 2,093 children, out of whom 494 males and 562 females are supposed to survive at 30.   As one-eighth of the males remain unmarried, 434 males

---

[1] *Ricerche sull' antica e moderna popolazione della città di Firenze per mezzo di registri di Battisterio di S. Giovanni dal 1451 al 1774*, Firenze, 1775.

[2] *Mémoire sur l'état de la population dans le pays de Vaud*, Yverdon, 1766.

are left, a number which again corresponds to 478 marriages, as about 10 per cent. marry twice. Twenty of these marriages are supposed to be sterile—a very low proportion according to modern experience, though Muret maintains that it is probably too high. There being 458 marriages with issue and the number of children being on an average 5·58 for each mother (375 mothers having as above mentioned 2,093 children), the new generation will count 2,556. He finds (not quite correctly) an increase of 23 per cent. compared to the previous generation born thirty years ago. Euler's calculations show a doubling in 110–12 years, not far from the above-mentioned 120 years.

As to life-tables, Muret does not go beyond the distribution per mille of the observed deaths as to age. In discussing the material he shows, however, much common sense, for instance, concerning the connection between the mean age at death and the birth-rate.

Though *America*, as we have seen, presented fairly good conditions for an evolution of vital statistics, contributions to any research were somewhat scanty. A rather interesting attempt at measuring mortality in an increasing population was made by Wigglesworth[1] on the supposition that the number of deaths was half the number of births. Unfortunately he does not explain his method sufficiently, so that it is hardly possible to reconstruct his calculations.

In *England*, as in America, observations on vital statistics continued to be made in this period. Richard Price's contributions were mentioned above. Two medical men may be cited, viz. Haygarth and Heysham. The former took particular interest in the problem of inoculation against small-pox; moreover, he issued a report on the mortality in Chester.[2] Heysham, who

---

[1] "A Table showing the Probability of the Duration, the Decrement and the Expectation of Life, in the States of Massachusetts and New Hampshire formed from sixty-two Bills of mortality . . . in the year 1789" (*Mem. American Acad.*, II, Part I, Boston, 1793).

[2] Westergaard, *Die Lehre von der Mortalität und Morbilität*, 2nd Ed., 1901, p. 58.

seems to have been a very original and energetic man, was also interested in the problem of small-pox, but his name is particularly connected with the famous *Carlisle-table*.[1]   For several years Heysham prepared lists of marriages, births, diseases, etc., in Carlisle, at that time rather an insignificant town with about 200 deaths yearly. A local census was taken in 1780, and again in 1787. Heysham was, however, not satisfied with the latter and undertook a private census himself.   He found the official census had omitted about 6 per cent. of the population. Many years after, his material came into the hands of J. Milne, who constructed a life table on the census results and the lists of deaths for 1779–87, after a very careful revision of the observations, based on a long correspondence with Heysham.[2]   This clearly shows how difficult it was in those days to get reliable statistical material.

On the *Continent* several investigations regarding *widows' funds* may be quoted, a subject which in the eighteenth century claimed much attention.   But it must be admitted that the actual results were on the whole trifling.   One of these authors was Kritter, who shows little originality with regard to statistical investigation.[3]   Like the Danish mathematician, Tetens, he tried to find the difference between the mortality of males and females.   As will be mentioned below, the latter deserved credit for his work in connection with life-insurance and theoretical statistics, but did not contribute much to mortality statistics.   In a great work, published 1785–6,[4] he discusses the mortality experience at hand.   He accepts Süssmilch's table, with Baumann's

---

[1] Henry Lonsdale, *The Life of John Heysham, M.D., and his Correspondence with Mr. Joshua Milne, relative to the Carlisle Bills of Mortality*, London, 1870.

[2] J. Milne, *A Treatise on the Valuation of Annuities and Assurances on Lives and Survivorship*, London, 1815.

[3] E.g. "Sammlung wichtiger Erfahrungen bey den zu Grunde gegangenen Wittwen- und Waisen-Cassen, Leipzig, 1780.   Untersuchung des Unterscheides der Sterblichkeit der Männer und der Frauen von gleichem Alter," *Göttingisches Magazin der Wissenschaften u. Litteratur*, B. II, 1781; confer Westergaard, l.c., pp. 56–7 and 290.

[4] *Einleitung zur Berechnung der Leibrenten und Anwartschaften*, I–II, Leipzig, 1785–6.

corrections, as sufficiently exact for Northern Europe, even though he finds that further corrections are possible. He is well aware that migrations may have influence, but on the other hand he warns against too much altering of the original observations (l.c., 1, p. 77). If the lists are correct, then the numbers in the mortality table can be considered as facts; if we alter these facts we must give reasons for the supposition we have made. He recommends, however, Price's Northampton table, which he finds classic.

34. In many respects statistical observations in those days were treated with critical sense, but in other fields much had yet to be learnt, the data often being taken as correct in spite of evident defects of the material. This may particularly be observed in the chapter of statistics which deals with *longevity* ("Macrobiotics"). The authors writing on this subject were often amateurs with regard to statistics, not having availed themselves of the latest progress.

Very naturally the question arose as to the utmost limits of human life. Instead of trying to find how a certain generation would gradually die out, the question was what age a human being could reach under the most favourable circumstances. This problem required less technical insight than the calculation of a life-table, but on the other hand observations had to be sifted with the utmost care if the efforts to solve the problem were not to be useless. Several authors touched on this problem, for instance Süssmilch. The famous scientist A. v. Haller may also be quoted here; he discusses longevity in his physiology.[1] In England, James Easton published a long list of "macrobiots",[2] and in America Will Barton discussed the problem.[3] One of the most

[1] A. v. Haller, *Elementa Physiologiæ Corporis Humani*, Tome VIII, 2nd ed., Lausanne, 1778.

[2] J. Easton, *Health and Longevity as exemplified in the Lives of Six hundred and Twenty-three Persons deceased in various Parts of the Globe remarkable for having passed the Age of a Century*, Salisbury, 1799 (new ed., 1823).

[3] W. Barton, "Observations on the probabilities of the Duration of Human Life and the progress of Population in the United States of America" (*Transact. of the American Phil. Soc.*, III, 1793).

characteristic contributions of this kind was C. W. Hufeland's book on the art of living long.[1]　This book enjoyed a very great reputation, which may have been well deserved as far as the medical advice which it contained is concerned, whereas the statistical value is very small. The observations on "macrobiots" are uncritical, and his calculations as to the frequency of deaths from various causes are extremely vague (l.c., pp. 366–7).　He also gives a mortality table (p. 214), which he describes as being based on experience, though without quoting its origin.　It seems to be calculated freely from Graunt's table.　According to this table, out of 100 newly-born, 50 will be alive at 10 years of age, 14 at 40 years, and 6 at 60 years.

[1] Hufeland, *Die Kunst das menschliche Leben zu verlängern*, 1st ed., Jena, 1797.

CHAPTER IX

PROGRESS OF THEORY AT THE CLOSE OF THE EIGHTEENTH CENTURY

35. Among the famous mathematicians of the Bernoulli family Daniel Bernoulli (1700–82) has just claims to be remembered in a history of statistics. The problem of inoculation against small-pox has been mentioned above. Dan Bernoulli added a weighty contribution to the discussion.[1] The material for his investigations was by no means faultless. As the basis for his calculations he used, for instance, Halley's table, and here he even misunderstood the numbers, supposing that the initial number 1,000 corresponded to the age 1 year. In order to get back to 0 year he adds, rather arbitrarily, 300 to the number of births, thus beginning with 1,300. He wanted to know the frequency of attacks of small-pox as well as their lethality (the rate of mortality among persons who have the disease). He also supposed the lethality as well as the frequency of attacks to be constant throughout life; both these qualities are estimated at one-eighth, one out of eight thus being yearly attacked, and one-eighth of these succumbing. All this may be doubted, but the merit of his paper is the masterly way in which he treated the theoretical problem. At first sight it is very complicated. Taking, for instance, the year as unity, we may ask how many persons of a given age contracted the disease in that year, how many recovered, and how many again of these died from other

---

[1] D. Bernoulli, "Essai d'une nouvelle analyse de la mortalité causée par la petite vérole et les avantages de l'inoculation pour la prévenir" (*Histoire de l'Acad. royale des sciences, année* 1760, *avec des mémoires de mathématique et de physique pour la même année,* Paris, 1766).

causes in the course of the year concerned. This Bernoulli simplifies by using *infinitely small units* of time. Dealing with a single moment he has only to find the probability of a person dying in that moment, without being embarrassed by future events. Using the force or intensity of mortality as the stepping-stone for further calculations, he obtains elegant formulæ, the problem being reduced to the solution of a simple differential equation. Having solved this question, he can easily find the relation between the number of persons—surviving according to the life-table which has been chosen—who have never had small-pox, and how many have had an attack but have recovered.

The paper concludes with some tables with numerical results. According to his supposition, out of 100 persons surviving at the age of 24 only 5–6 would never have had the disease. This may be wrong but, in regard to method, his investigations represent a great progress, and only correct observations will be required in order to reach reliable results. By his invention of the *continuous method* he opened up a wide field for statistical investigations. Unfortunately his method was not generally understood; it was not until late in the nineteenth century that it was thoroughly appreciated.

Soon after sending his paper to the academy he was vigorously attacked by d'Alembert,[1] who was, however, not very just in his criticism. D'Alembert is right in maintaining that the individual member of a population will necessarily look upon inoculation with other feelings than the State (Society). It may be of profit to the latter if inoculation adds some years to the expectation of life, but the private citizen when considering profit and gain will ask whether there is a risk of dying as a result of the inoculation. Daniel Bernoulli seems to have seen this clearly himself, and in other respects d'Alembert's objections carry little weight.

A single mathematician may be mentioned who

---

[1] d'Alembert, "Sur l'application du Calcul des Probabilités à l'inoculation de la petite Vérole" (*Opuscules Mathématiques*, II, 1761).

understood the method thoroughly, viz. Duvillard
(1755–1832). He had in 1787 published a book
on interest.[1]   At the beginning of the nineteenth century he was in the public service (population statistics).
The newly introduced vaccination against small-pox gave
him quite naturally an impulse to take up the problem
which D. Bernoulli had treated; the result was a remarkable volume which unfortunately became very little
known.[2]   He speaks with veneration of D. Bernoulli's
work and follows his methods, but, having more complete
observations at hand, he is enabled to go more into
detail.   It may be objected that his mortality-table is not
essentially better than the table which D. Bernoulli used.
It was constructed on about 101,000 deaths on the
supposition of a stationary population.   He was himself
of the opinion that the table was a good expression of the
state of health in France before the Revolution.   But it
was not his principal aim to calculate a life-table but
only to obtain means for a comparison of the effects of
vaccination, and in this respect it would probably serve
fairly well.   Unfortunately, several authors who afterwards criticised Duvillard have only had this table in
view, without thinking of the problem under discussion.
As stated, his book is, on the whole, very little known,
only very few copies seem to be preserved.

As to the occurrence of, and death-rate from, small-
pox, various observations with distribution according to
age were at his disposal (Geneva, the Hague, Berlin),
and of these he took an average which he used for his
calculations.

His book contains a considerable number of tables.
The population corresponding to the life-table is divided
into various classes: persons who it is assumed will never
be attacked by small-pox, who will recover from an
attack, who will die from the disease, etc., on the whole

[1] *Recherches sur les rentes, les emprunts et les remboursemens* . . . Par M.
Du Villard, Genève, 1787.

[2] E. E. Duvillard (du Léman), *Analyse et tableaux de l'influence de la petite
vérole, et de celle qu'un préservatif tel que la vaccine peut avoir sur la population et la longévité,* Paris, 1806.

following the line of Bernoulli. It may be difficult for readers in our days to judge the assumptions with absolute clearness. He found that out of 100 persons aged 30 years, only 3 had not had small-pox, and that two-thirds of all newly-born will be attacked. The mortality of persons suffering from small-pox was, according to his calculations, no less than one-third at the age of one year, 3 per cent. at 10 years. If small-pox disappeared, a new-born child would on an average gain $3\frac{1}{2}$ years in expectation of life.

Even if Duvillard's work had been generally known it is doubtful whether his results would have been accepted. The leading economists of those days did not believe in great changes in mortality and expectation of life, such as Duvillard did. If economic conditions did not improve they would expect that somehow or other the gain would be out-balanced by losses, so that the result would be reduced to zero.

Strictly speaking, Duvillard's book should have been treated in a following chapter, but it was natural to mention it here on account of its close relation to D. Bernoulli's treatise.

36. Although the continuous method was very little noticed, another important progress may be mentioned though only indirectly connected with the history of statistics. J. N. Tetens has, in his above-quoted work, used a method of great actuarial usefulness, what in modern technical terms may be called the method of D_x and N_x columns. We have seen that the mathematicians of the eighteenth century, such as de Moivre and Simpson, had thoroughly grasped the problem of finding the present value of sums payable with a certain probability at given dates. But if an assurance embraced many payments at various dates, as, for instance, an annuity on life, the calculation might require much time. De Moivre tried, as mentioned above, to find a short cut by his hypothesis on the law of mortality, but his solution was not general enough. The problem is analogous to that of finding the value with compound interest at a

given date of several instalments in a savings-bank. The greater part of the work can be done in advance, by calculating the value of each instalment at a previously fixed date, for instance, the beginning of the current calendar year, the problem thus being reduced to finding the value of the aggregate of these sums at another date. Just the same can be done in life insurance, choosing, for instance, the age 0 as starting-point.[1] This invention was in fact a real Columbus-egg. Tetens' method was, however, not much known. English actuaries, to whom Tetens' work was unknown, even invented the method many years after.

Credit is also due to Tetens for making an energetic attempt at finding the risk incurred by a life-insurance society. He shows that the total risk of a widow-fund increased as the square-root of the number of members. He refers to de Moivre, but he reached his result by an original process.

Finally, Tetens was one of the first to use the method of *expected deaths* in order to compare the mortality in a widow-fund with that of a certain mortality-table—in this case Süssmilch's table (l.c., II, pp. 1 *sq.*). His method is, however, a little unpractical though perfectly clear. The period of observations is 1767–83, and he includes the members who enter in the same half-year in a separate group, asking how many of these would be expected to survive in 1783 according to the table, and comparing the result with that actually experienced. The method does not enable him to find the mortality in each age of life. For females, the main result was 743 actual and 1,048 expected deaths.

37. If it was difficult to gauge the population, there were still greater obstacles for getting correct figures concerning *economic statistics*. Here and there progress was made. Thus the United States had official statistics

---

[1] l_x being the number of survivors at the age x and r the rate of interest, we have $D_x = l_x(1 + r)^{-x}$ and $N_x = \Sigma_x.D_x$ and N_x being calculated beforehand, the calculation of annuities and other insurances can be simplified very much.

of foreign commerce compiled each year beginning with
1789.   Agriculture was not much dealt with statistically
although occasionally details may occur in the writings,
for instance, figures showing the number of cattle and
horses in Alsace in Boulainvilliers' above-quoted work,
but generally very little was known even much later in
the century, in spite of the great interest aroused by the
physiocrats.   At the close of the century, however, a
remarkable little book appeared which made notable
attempts in this direction.   The sixty-four pages contain
contributions from four authors, first of all the famous
chemist Lavoisier (born 1743, decapitated 1794) and the
mathematician Lagrange (1736–1813).[1]

Lavoisier had already in 1784 begun his investigations
in this field.   He does not consider his report as complete, but the Assemblée Nationale ordered its publication.   He refers to Moheau and de la Michaudière
(Messance) regarding the number of inhabitants, and
without going into detail as to how he has arrived at the
result he gives an estimate of the total population at about
25 millions, which he again distributes according to
various classes, supposing 8 millions to live in towns,
whereas $2\frac{1}{2}$ millions are getting their living from vine-
yards, etc.   His principal assumption is that the consumption per head of certain necessities of life is nearly
constant within each class, and after having found plausible support from statistics on consumption in Paris, he
finds that the yearly consumption of wheat, rye and barley,
including seed, was 14,000 millions of "livres pesant,"
and neglecting export and import, he considers this
quantity as the yearly average produce.   He now pro-

---

[1] *Collection de divers ouvrages d'arithmétique politique par Lavoisier, Delagrange et autres*, Paris an IV.   The four papers are the following: Lavoisier,
"Résultats extraits d'un ouvrage intitulé"; "De la richesse territoriale du
Royaume de France, 1791"; "Réflexions d'un citoyen propriétaire sur l'étendue
de la contribution foncière et sa proportion avec le produit net territorial
converti en argent" (author unknown), 1792.

Le citoyen Delagrange, *Essai d'arithmétique politique sur les premiers besoins
de l'intérieur de la République*.

Antoine Diannyère, *Preuves arithmétiques de la nécessité d'encourager l'agriculture*.

ceeds to question how many ploughs and how much cultivated area will be sufficient to produce this quantity. Various investigations combined with his own practical experience led him to the result that a plough drawn by horses will suffice for 27,500 livres pesant, whereas a plough drawn by oxen cannot give more than 10,000. One plough will again correspond to 90 and 60 arpents respectively.   Knowing, further, in which part of the country it is customary to use horses, and where oxen are prevalent, he can find the number of animals occupied in this labour, but various groups will have to be added, thus for horses, the number used in agriculture where oxen are prevalent, also the numbers in the towns, and in the army.   Here he is obliged to use various estimates.   He admits himself that this part of his work is very hypothetical,[1] but he hopes by further observations to get more correct results.   The total number of horses will, on his estimate, be 1,781,000, whereas there are supposed to be 7,089,000 cattle, 20 million sheep and 4 million pigs.[2]

As the main result with regard to the area, Lavoisier finds that more than one-third of the total area was left uncultivated, and he seems himself a little surprised. He proposed a great institution—a statistical department, where all particulars concerning agriculture, commerce and population, etc., might be collected.   He maintained that there was only one country where an institution of this kind could succeed, viz. France; it depended only on the will of the Assemblée Nationale.[3] Here the whole economic result of the French agriculture —in goods as well as in net income, should be calculated. He makes a characteristic remark reminding all of the

---

[1] Cette partie de mon travail est, comme l'on voit, fort hypothétique.

[2] Comparing with modern statistics we find (1926) 7 horses per 100 inhabitants; the same result has Lavoisier, cattle 35 against 28 according to his estimate, sheep 26 and 80 respectively, swine 14 and 16.   Taking into consideration the great changes in agriculture which have taken place, Lavoisier's results seem to be in rather good harmony with modern figures.

[3] "l'Assemblée nationale n'a que le désirer et le vouloir"; cf. above, art. 21, Fénélon.

physiocrats (among whom he quotes Quesnay as well as Turgot) that a report of this kind might contain in a few pages the whole science of political economy, or rather this science would cease to exist, the problems being so easily solved that no disagreement could be possible.[1]

A following paper, by an anonymous author, continues Lavoisier's estimates of the produce of the French agriculture and tries to give a more optimistic impression, whereas Lagrange in the next paper deals with the consumption in the whole empire, particularly investigating the consumption in the army. The last article, by Diannyère contains an attempt at proving the correlation between prices of corn and mortality. On the whole, all these articles are worthy supplements to the weighty treatise by Lavoisier, though they cannot by any means compare with it, but the small volume gives an insight into the difficulties with which political arithmeticians still had to contend at the close of the eighteenth century, at the same time showing the ingenuity which enabled Lavoisier to draw outlines of economic statistics.

[1] "Un travail de cette nature contiendroit, en un petit nombre de pages toute la science de l'économie politique, ou plutôt cette science cesseroit d'en être une; car les résultats en seroient si clairs, si palpables; les différentes questions qu'on pourroit faire, seroient si faciles à resoudre, qu'il ne pourroit plus y avoir de diversité d'opinion."

CHAPTER X

CALCULUS OF PROBABILITY AND STATISTICS

38. IT is outside the plan of this treatise to describe the purely mathematical evolution of calculus of probability; I shall only mention what has been important for the progress of statistics.

Games of dice or cards early attracted the attention of mathematicians. Complicated problems came under discussion, often claiming high intellectual powers and clear thought. The solution of such problems laid the foundation-stone of the Calculus of Probability.

The beginning seems to have been made in Italy, where already in the sixteenth century Cardan (who was said to be an inveterate gambler himself) wrote a small paper: *De Ludo aleæ*, on such problems; later Galileo took up similar subjects.[1]

It is not difficult for a modern statistician to see how promising for the future of statistics investigations of this kind would be. Dealing as they did with the frequency of events, even though of an apparently insignificant nature, they might gradually be extended to more important realms, as, for instance, rates of mortality. Two problems especially would claim attention: *first*, if a probability of an event were known, how would the result of a certain number of trials—though deviating more or less from the known probability—lie between certain limits, and *further*, the probability of the event being

---

[1] Todhunter, *A History of the Mathematical Theory of Probability*, 1865, pp. 3 *sq.* Further information concerning the origin and history of this chapter of mathematics can be found in Cantor, *Vorlesungen über Geschichte der Mathematik*, II–III, 2nd ed., 1900–1; IV, 1908; and J. M. Keynes, *A Treatise on Probability*, 1921.

unknown, how near the truth would a certain number of trials bring us? The number of male births observed being, for instance, 103 for each 100 female births, would it be possible to say within what limits the normal ratio would lie? Here a co-operation between abstract and numerical experiments was required, as it would be necessary to know whether the observations in various fields of statistical investigations would correspond to theory, and what were the conditions for deviations from the laws to which abstract thinking had led.

Unfortunately the immediate influence on political arithmetic was very small. Investigations in the Calculus of Probability were mostly of a purely abstract nature. Generally the mathematicians interested in the new science quite dogmatically supposed that experience would harmonise with theory. By their investigations they promoted pure mathematics, but it took a very long time before the calculus of probability came into closer contact with statistics. It may even be asserted that only our own generation has derived full profit from a co-operation of this kind.

Isolated cases of experimental observations exist, but they are very rare, although undoubtedly many gamblers made several practical observations on the frequency of various combinations, leading them to form their own theory of how to play with advantage on the basis of this experience, and Galileo in an undated paper gave a reply to a question of this kind. Gamblers had found that three dice more frequently show the number 10 than the number 9. Galileo shows that out of 216 possible cases, 27 are favourable for the appearance of the number 10 whereas only 25 throws will show the number 9. In a nutshell we have here one of the chief problems of the calculus of probability. But, as remarked, cases of this kind are rare, and generally the ways of the new science were separate from those of political arithmetic.

Though the cradle of this science stood in Italy, French mathematicians deserve merit for the first earnest efforts to master such problems. Here we meet with names such as Pascal (1623–62) and Fermat (1601–65), who in the middle of the seventeenth century corresponded with each other about various problems which a certain Chevalier (later Marquis) de Méré had suggested to Pascal.[1] De Méré had no mathematical training, but he appears to have taken much interest in the problems concerned.

Pascal deals particularly with the so-called *problem of points*. Two players, whom Pascal supposes equally skilful, agree that he who first gets a certain number of points, for instance 3, has won. The first player is further supposed to have won 2 points in the course of the game, the second only 1. They now wish to stop, and the problem is how to divide the stakes, 32 pistoles (Louis d'or), for each player. Pascal solves the problem by considering the position of the second player. If he wins one point the next time, then the two players will stand equally, the first player being therefore entitled to 32. But if the second player loses in playing for this point, the first player has won 3 points and is consequently entitled to the whole stake of 64. It will therefore be just to give the first player 48 pistoles, the second one only getting 16. In the modern terms of calculus of probability this solution may be expressed as follows : the second player has a probability one quarter of winning, for it will be necessary for him to win a point twice successively, whereas the first player has a probability three-quarters, so that the stake will have to be divided as stated into 48 and 16.

Pascal now generalises the problem, and Fermat co-operates ingeniously with him in deepening the questions before them, so that it may truly be said that these discussions opened a new field for human thought. About the same time Chr. Huygens took up the matter in a treatise about 1657, *De Ratiociniis in Ludo Aleæ*, in which he solved various problems, for instance, of points.

---

[1] *Œuvres de Blaise Pascal*, publiées par Léon Brunschvigg et Pierre Boutroux, III, 1908, pp. 369–431; here also Fermat's letters will be found.

His treatise, over a long period, may be considered the best account of the subject.[1]

39. But by far more important were the contributions to calculus of probability for which we are indebted to Jacob Bernoulli (1654–1705). He began his studies on this subject as early as 1685 or before, but his *Ars Conjectandi* did not appear till 1713, eight years after the author's death. Unfortunately he left his work unfinished. Various attempts were made to prevail upon other mathematicians to finish the work, but in vain. But even as it was, Bernoulli's unfinished work is one of the most brilliant contributions to the literature of the calculus of probability.

Ars Conjectandi is divided into four parts. In the first he reprints Huygens' essay with comments. Then follows a theory of permutations and combinations. The third part of the volume deals with various problems of games. In the fourth, unfinished, part, it was his plan to deal with applications of the theory to morals and economics. This part contains the problem which bears his name. He proves that by taking the number of trials large enough the probability that the result will lie between certain limits will be as great as we wish. As an example he chooses $\frac{30}{50}$ as the probability of the happening of an event, and he now wishes to find the probability that the ratio of the number of times in which the event happens, to the whole number of times, will lie between $\frac{31}{50}$ and $\frac{29}{50}$. He arrives at the result that 25,550 trials would give the probability 0·999. By adding 5,708 trials to the number, thus basing on 31,258, we get 0·9999, whereas another 5,708 trials, altogether 36,966, would give 0.99999. Though these results point in the right direction, they are not in actual harmony with the results which modern instruments would give, and Bernoulli's demonstration, exact as it was, was later superseded by a more simple process. But at all events this achievement was a masterpiece. If statisticians had followed the track, they might at a very early epoch of the

---

[1] Todhunter, l.c., p. 25.

evolution of political arithmetic have reached the means of controlling and criticising many results which they were now only able to treat in a very vague manner. Bernoulli himself was well aware of the range of his theorem. It leads him to interesting meditations connected with Plato's philosophy. By continuing the trials *ad infinitum* we should at last be enabled to calculate everything with certainty and to see the order in chance.[1]

Several years after, Abraham de Moivre took up the thread in a most ingenious way. He had taken an interest in the calculus of probability before *Ars Conjectandi* appeared, publishing a memoir on this subject (*De Mensura Sortis*) in the *Philosophical Transactions*, 1711. Seven years after he published his work, *The Doctrine of Chances*, of which a second edition appeared in 1738 and a third one in 1756, two years after his death. Bernoulli's problem is not touched on in the first edition, but not many years after he found a solution, and in 1733 he wrote a small paper on this subject, which he communicated privately to some friends and later incorporated in the second edition of his Doctrine of Chances. In this paper he states that "a dozen years or more" before he found the ratio which the middle term of the binomial $(1 + 1)^n$, where n is a large number, bears to the sum of all the terms; later he extended his calculations to terms distant from the middle term, and by means of the formula of his "worthy and learned friend, Mr. James Stirling," simplified the solution.

De Moivre gives a very instructive example of this problem, which he calls the hardest that can be set on the subject of chance. The number of trials is 3,600, and the probability of the event is $\frac{1}{2}$; he finds that the probability of the event neither appearing more often than 1,830 times, nor more rarely than 1,770 times, will be 0·682688. In modern terms we should express this result so, that the mean error is $\sqrt{3600 \cdot \frac{1}{2} \cdot \frac{1}{2}} = 30$, and the probability that the deviation from the middle shall

____
[1] *Ars Conjectandi*, Pars Quarta, p. 239.

be less than the mean error is 0·683, which corresponds very well with de Moivre's solution.

De Moivre extends his solution to the case where the probability of the event differs from $\frac{1}{2}$; he also shows the effect of extending the limits, for instance, by doubling or trebling them, "for in the first Case the Odds will become 21 to 1, and in the second 369 to 1, and still be vastly greater if we were to quadruple them." These results are quite in harmony with the values found by means of the usual table of the exponential formula.

De Moivre was well aware of the importance of these results and their applicability to practical experience, for instance, as to the ratio of male and female births. Thus he makes the following remark :

"As upon the Supposition of a certain determinate Law according to which any Event is to happen, we demonstrate that the Ratio of Happenings will continually approach to that Law, as the Experiments or Observations are multiplied: so *conversely* if from numberless Observations we find the Ratio of the Events to converge to a determinate quantity . . ., then we conclude that this Ratio expresses the determinate Law according to which the Event is to happen."

And, just as Süssmilch, he draws the conclusion that there is a "great Maker and Governor of all," and he maintains that although "Chance produces Irregularities, still the Odds will be infinitely great, that in process of Time these Irregularities will bear no proportion to the recurrency of that Order which naturally results from Original Design." But there is the curious difference between Süssmilch and de Moivre that the former arrives at his conclusion by studying practical experience, whereas de Moivre confines himself to abstract reasoning, and even though de Moivre slightly touches statistical observations—as remarked above—there was not in fact any bridge yet between political arithmetic and the calculus of probability. We have seen that de Moivre contributed to political arithmetic through his *Annuities on Lives*, but even here we miss the influence of these more intricate problems. But at all events this cannot deprive him of the honour of having solved a problem of very great importance to statistical investigations.

40. The fact that no fertile co-operation materialised between the mathematicians who took up the calculus of probability and the political arithmeticians can be attributed to a certain lack of clearness in the definitions. It is not unconceivable that even Jacob Bernoulli would have had difficulties with regard to the applications to Morals and Economy which he had planned as the finishing part of his *Ars Conjectandi*.

We meet this lack of clearness in the so-called *Petersburg Problem*, which owes its name to the fact that it first appeared in the commentaries of the Petersburg Academy.[1]

The problem is related to those systems which so many gamblers in our days form in order to get an infallible chance to win: if they lose in the first turn, they increase their stake in the second turn, so that they realise a profit if they win this time; if not, they increase the stake again, etc. Their mistake is simply that they do not take into consideration the chance that they may never win.

The Petersburg Problem may be expressed in the following way: A player tosses a coin in the air; if head appears he is entitled to a shilling from the other player, if it is tail, he goes on and gets 2 shillings if head appears in the next throw. If he loses the second time he plays again and is entitled to 4 shillings; if he wins in this turn, etc. The puzzle is that the expectation of the first player is evidently $\frac{1}{2}$ for each throw, and consequently infinite if the game is to be continued as long as tail appears, but common sense would reply that no prudent man would pay even a small sum of money to enter into a game of this kind.

---

[1] Daniel Bernoulli, "Specimen Theoriæ Novæ de Mensura Sortis," *Commentarii Academiæ Imperialis Petropolitanæ*, Vol. V (for the years 1730–1, published 1738). It appears from a letter inserted in the Memoir that problems of this kind were already discussed in 1713.

Various solutions were proposed, more or less appealing to common sense. Buffon,[1] who says that he met the problem in 1730, answers that there would not be time during a whole life for playing more than a certain number of games. Moreover, he quite arbitrarily proposes that any chance smaller than 0·0001 should be considered absolutely zero. Cramer, in a letter quoted in Daniel Bernoulli's memoir, suggests that the pleasure derivable from a sum of money may be taken to vary as the square root of the amount. He also proposes to consider all sums greater than 2^{24} as practically equal.

The most radical solution was proposed by d'Alembert. Constantly opposed as he was to the results of the calculus of probability, he maintains that if tail arrives three times in succession it is more likely that head will be shown next time than tail, and the oftener tail arrives successively the greater will be the chance of head at the next throw.[2] In this way he can easily find a solution which gives the player a finite expectation. Here he is in accordance with many gamblers, for it is a very popular idea that the probability changes after the results of the previous tossings, and in the above-mentioned systems this idea will often be the foundation-stone. The Danish poet and philosopher J. L. Heiberg has elucidated this in two monographs. Thus he describes two methods which are favourites among gamblers: Paroli and Martingale. According to the first method it is supposed that there is a momentary rule in the game, and the player supposes that this rule will hold good in the next turn. Martingale has the opposite basis; it is supposed that the momentary rule will soon be broken, that, for instance, tail will have more chance of appearing if the coin has shown head several times in succession.[3] It would not be difficult to gather a large collection of tracts

---

[1] Todhunter, l.c., p. 346.

[2] *Réflexions sur le Calcul des Probabilités*, Opuscules, II, 1761.

[3] J. L. Heiberg, *Der Zufall aus dem Gesichtspunkte der Logik betrachtet* (1825); *Nemesis* (1827).

and articles on the problem of how to win in games, based on ideas of this kind.[1]

On the other hand, the supporters of the theory of probability criticised d'Alembert's hypothesis energetically. Thus Euler declared it quite absurd. If it were true that a number in a lottery would appear more frequently in future, if it had not been drawn several times, then the same would hold good if the drawings took place a year or even a century later, and also if the drawings were made anywhere on the earth.[2] Curiously enough, d'Alembert himself suggests an experiment to confirm his theory; but he never seems to have realised this idea himself. It was, however, tried by Buffon,[3] who let a child toss a coin in the air. If tail was shown the coin was thrown again, etc. This experiment, repeated 2,048 times, quite supported the common theory. It is one of the very few cases from earlier times in which experiments were made in connection with the theory of probability.

As to the paradox of the Petersburg Problem, it must be remembered that one side of the question is to calculate the expectation concerning each throw, another the inducement to the player or the bank to enter into the game. In other words: not only the *expectation* should be calculated but also the *risk*, just as a life-office must have reserves to cover accidental deviations from the expected expenses. In this way it will be easy to conciliate theory with common sense.

41. So far the discussion of the Petersburg Problem is of great interest inasmuch as it gave the first impulse to the theory of Marginal Utility which in modern economy is of so great importance.[4] Daniel Bernoulli, in fact, gives the theory in a nutshell in the above-quoted memoir,

---

[1] For instance, Gardon, *Antipathies des 90 Nombres, Probabilités et Observations comparatives sur les loteries de France et de Bruxelles,* 1801.

[2] *Opuscula Analytica,* Vol. II, 1785 ("Solutio quorundam questionum difficiliorum in calcolo probabilium").

[3] *Essay d'Arithmétique Morale,* 1777.

[4] Compare, among others, Filippo Virgilii, "Speranza matematica e speranza morale," 1911 (*Estratto degli Studi Senesi,* Vol. XXVIII).

and as remarked above also, Cramer gives a hint in the same direction.   Bernoulli's theory points to the fact that a rich man will attach less value to a certain increment to his fortune than a poor man will, and he particularly considers the case that a small increment has a relative value inversely proportionate to the sum, thus arriving at a simple logarithmic formula for the total utility. Daniel Bernoulli calls this subjective value *emolumentum*; Laplace, adopting the theory, uses another term, viz. *la fortune morale*.[1]   This terminology is in so far less fortunate, as the "emolumentum" has nothing to do with ethics.

Daniel Bernoulli can justly be considered as the predecessor of the modern theory of value, but unfortunately he is led astray when trying to apply his theory.   Thus he will find the value of various possible incomes, each with its peculiar probability of being realised, for a person with a certain fortune.   This *emolumentum medium* he finds by adding the single emolumenta multiplied by their probabilities.   The problem is, however, much more complicated.   The marginal utility is a function of the original sum, and all the various profits and probabilities, and it has to be proved that this function is of the simple nature which Bernoulli presumes.   In fact, modern political economy still struggles hard to solve the problem.

That Bernoulli's solution cannot be accepted can easily be proved.[2]   One result of his calculations is, for instance, that any game is disadvantageous to the player, for he gets a smaller *fortune morale* than before he entered into the game.   But as it is not a question of morals it can be justly objected that it is impossible to explain why there are so many persons who, in spite of the loss they occur according to the theory, are still willing to

---

[1] "Recherches sur l'intégration des équations différentielles . . ." (*Mémoires de Mathématique et de Physique . . . par divers Savans*, Tome VII, Année 1773, published 1776).   Having exposed Dan Bernoulli's theory he adds, "Cette règle n'est cependant pas générale, mais elle peut servir dans un grand nombre de circonstances."

[2] Westergaard, *Den moralske Formue og det moralske Haab. Tidskrift fo Mathematik*, 1876.

play. The question is not whether they *ought* to play, but whether they actually *will* play.

Dan Bernoulli's ingenious theory failed to influence political economy. The time had not yet come for investigations of this kind, just as there was no bridge between the calculus of probability and political arithmetic, and so the theory of *la fortune morale* was left unnoticed among economists through several generations.

42. The theory of probability received weighty contributions from Laplace, who while still a young man wrote various interesting memoirs which later more or less became elements of his *Théorie analytique des Probabilités* (1812). This work gives a true picture of the state which the theory had reached as the result of the efforts of past centuries, an evolution in which Laplace himself took a conspicuous part.

The great problem which Jacob Bernoulli and de Moivre had touched, was also treated by Laplace, viz. the reliability of conclusions from numerical observations. In a memoir published in 1781[1] he discusses the observations in Paris for twenty-six years concerning male and female births, altogether 493,478, out of which 251,527 were males. The question is, for instance: what is the probability that the ratio of boys is greater than $\frac{1}{2}$? He finds that there is an extremely small chance for the ratio to be smaller than $\frac{1}{2}$. It would of course have been useful if he had proceeded one step further, asking between what limits the ratio actually lay. The mean error being 0·0007, it will be seen that the ratio is at least 0·507 and not higher than 0·512, so that the ratio can be correctly stated by using only two decimals (0·51).

Here also we can mention his attempt at finding the population of France which was discussed above (art. 30). It is an earnest endeavour to build the bridge between political arithmetic and the calculus of probability.

There is reason to refer to another problem treated by mathematicians in the eighteenth century as a chapter of the calculus of probability. The question would natur-

---

[1] "Mémoire sur les Probabilités" (*Histoire de l'Académie* for 1778).

ally arise in Physics, Astronomy and Geodesy, of how to deal with discordant results of observations. In its simplest form the problem was to find a plausible value of a single element which had been measured several times with differing results. Under the supposition that a certain law rules the phenomena, as, for instance, the movement of a planet, it would further be the task to calculate the orbit from a certain number of observations which were supposed to deviate more or less from the true values, and the various more or less complicated questions arising here would claim a solution.

The problem was mooted early in the century by the English mathematician Cotes; later, several others became interested in it, as Simpson, Legendre, Lagrange and Daniel Bernoulli.[1] Lastly, Gauss and Laplace contributed to the solution of the problem. The discussion led to the method of *least squares*, which was justified by the supposed distribution of the observations according to the binomial law. Thus the problem was intimately connected with the theory of probability, and the evolution in the nineteenth century proved the applicability of the method in statistical investigations, for instance, by the adjustment of the frequently very irregular values of rates of mortality in a life-table calculated from practical observations.

43. Before turning to the progress of statistics in the nineteenth century it may be worth while to ask what had hitherto been reached as the results of the efforts in the seventeenth and eighteenth centuries and what remained to be done.

First of all, there was a lack of *statistical observations.* The observations which had been collected, partly through private efforts, were as a rule only fragmentary. Sweden had made an excellent start with regard to official statistics, but on the whole there was too much diffidence in the population as regards enumerations; it disappeared

---

[1] E.g. an investigation by Simpson. "An attempt to show the Advantage arising by Taking the Mean of a Number of Observations in Practical Astronomy," 1757.

gradually in the nineteenth century, opening the way for a fertile co-operation between the authorities and the population in compiling the mass of statistical information which sometimes even caused an "embarras de richesses." The growth of *official statistics* under this increasing co-operation is perhaps the most interesting feature in the history of statistics in the nineteenth century.

Wrong or imperfect methods were abundant. They were quite natural during the first period of evolution, and they proved deeply rooted in the following generations; it was rather a hard task to do away with them, so that investigations could be based on correct principles. It is sufficient here to recall life-tables which had in many cases so evident defects, in spite of which they often enjoyed general confidence.

Lastly, the *calculus of probability* had to be adapted for statistical investigations. It was necessary to know whether statistical data followed the binomial law, and what were the conditions for the application of this law, or whether the observations obeyed other laws which consequently required to be explored. This was absolutely necessary for it to be possible to find tests for the checking of the conclusions from the statistical material.

The political arithmeticians knew a great deal: for instance, about balance of sexes, about mortality under various conditions, etc., but their knowledge was often vague and uncertain, frequently based more on common sense than on exact investigations, and they were therefore often unable to meet criticism. The task of the nineteenth century would naturally be to find out *what was really known*, and gradually to add soundly founded knowledge to that which had already been acquired. This was perhaps more fatiguing and less satisfying than the work of the pioneers, but it was at all events highly necessary.

CHAPTER XI

OFFICIAL STATISTICS AT THE BEGINNING OF THE NINETEENTH CENTURY

44. CONDITIONS for a fertile evolution of statistics were in many respects favourable in the nineteenth century. The modern governmental machinery made it easier to compile statistical observations—particularly in countries where a free constitution had been introduced—and gradually the prejudice against publication of statistical documents disappeared. Many treasures hitherto hidden in governmental archives were now made accessible to the public. Not only did official statistics profit by these altered conditions, but their influence was felt in the whole of statistical literature, which everywhere received the stamp of modern scientific thought.

This change did not, however, take place at once. In the first three decades of the century political reaction was, in many countries, a serious obstacle to official statistics, and, on the whole, statistical literature was in many respects poor, and akin to that of past centuries. An exception was actuarial science which counted several valuable contributions. But the years around 1830 were a turning-point. The thirties are an epoch of revolution which introduced a long period of uninterrupted growth. It would, however, not be just to look on the first three decades of the nineteenth century as a barren and unfertile chapter of the history of statistics; as we shall see, real progress was made, particularly in connection with life insurance, and this period had in any case a hidden growth which in due time would show visible results. But in several respects it will be natural to treat these three decades as a continuation of the age of the political arithmetic.

45. In *France* an apparently promising step was taken in 1800 by the formation of a Bureau de la Statistique Générale. Further, an Act was passed in this year, prescribing a general enumeration of the population. This census, however, appears to have been a great failure. Lucien Bonaparte, who was then Minister of the Interior, wished to get the results within two months, but it took in fact two years.[1] Laplace's above-described experiment of an estimate of the population by means of the birth-statistics testifies to the diffidence with which the census was regarded, but this is not the only proof. Thus, in a letter from the prefect of Morbihan, it is maintained that there are communes which have raised their population to 4,000 souls, though in fact they had only 2,000 inhabitants.[2] Nor was the following census a success (ordered by a circular of November, 1805, and taken in 1806). Later, a general census was prescribed for the year 1821, and again in 1826 and 1831, but as Faure remarks, it is probable that the official tables were obtained by a simple estimate. The Bureau de Statistique was suppressed in 1812, and it was not re-established till 1833.

A statistical society (Société de Statistique de France) which had been founded by Ballois in 1803, had the same fate of suppression. It disappeared already in 1806, and with few exceptions it was hereafter forbidden to publish statistical reports. Ballois also founded a statistical journal, *Annales de Statistique*, under the auspices of the Government, the first number being published in 1802, but this journal only lived for two years.

Official statistics had thus rather hard times in France, in spite of the promising beginning early in the century. Some cities, however, particularly *Paris*, cultivated municipal statistics. In *Paris* a general census was taken in 1817 and the famous scientist Fourier made contributions to the reports, which will be mentioned below. It may

[1] Faure, l.c., pp. 285-7.
[2] *Annales de Statistique*, Seconde Année, Julliet, 1803.

be added that annual statistical publications, edited by the Ministry of Justice, appeared from 1827; about the same time this Ministry got its own Bureau de Statistique.

A comparatively rich statistical literature saw light in the few years early in the century, before the reaction. Duvillard's above-mentioned investigations, 1806, as to the influence of vaccination on small-pox, were sufficiently neutral to escape suppression. Ballois was active as editor of the statistical annals. He died, however, in 1803, only 25 years old, the annals hereafter being edited by a *société de gens de lettres* till they were discontinued. These annals contain mostly descriptions of the various departments of France, with relatively few numerical observations. From the same period a work in seven volumes can be quoted, published by Herbin, being a general statistical description of France and its colonies.[1] Among the co-operators was Peuchet, who wrote an introduction dealing with the history of statistics. It is, however, not very accurate as regards the statistical literature outside France. The same author published a *Statistique élémentaire de la France* (1805), containing what was at hand of statistical observations concerning France. With the exception of Duvillard's work these publications are now of relatively small interest.

Some statistical reports, concerning Paris, were, as mentioned above, written by Fourier (1768–1830).[2] In his advanced years he was in a difficult economic position, and the prefect of the Department of Seine, Chabrol, one of his former pupils, tried to help him by attaching him to municipal statistics. These investigations are, however, very little known, and they seem to have had no influence on the statistical literature of the following

---

[1] *Statistique générale et particulière de la France et de ses Colonies . . .* par une Société de gens de lettres et savans et publiée par P. E. Herbin, Paris, 1803.

[2] In *Recherches statistiques sur la Ville de Paris et le Département de la Seine: Notions générales, sur la population* (Vol. I, 1821); *Mémoire sur les résultats moyens deduits d'un grand nombre d'observations* (Vol. III, 1826); *Second Mémoire sur les résultats moyens et sur les erreurs de mesure* (Vol. IV, 1829); all these reports are anonymous.

period. It would naturally be expected that when a genius of his rank took up subjects of a statistical nature, he would have thrown new light on the problems. But even though these reports stand higher than most of those written by his contemporaries,[1] his investigations should perhaps be looked upon more as occasional work, than as work dictated by inner impulses. Fourier shows how a stationary population is constructed according to age, under the influence of the rates of mortality in various ages. Further, he touches upon the influence of migrations, though rather vaguely; he appears to have refrained from treating this problem more closely on account of the mathematical analysis which it required, and which he found it difficult to explain in an elementary form.

46. The expansion of the French Empire during the *Napoleonic Wars* brought various *German* territories under the influence of the French statistical machinery. Sometimes the practical results may have been even better than generally in France proper. This seems, for instance, to have been the case with the Departement du "Mont Tonnerre" (i.e. Donnersberg in Pfalz), where the census of 1801 was taken, and where several other subjects were treated, such as prices, education and agricultural produce.[2] But independently of this influence from *France*, statistics on the whole made progress in *Germany*. Much material had already been collected in the eighteenth century; for instance, in *Prussia* (under Frederic the Great), where the economist Leopold Krug (1770–1843) obtained access to the data, being thus enabled to publish a statistical description of *Prussia*,[3] and further a treatise on National wealth in the same country.[4] In the latter work he tried, in the spirit of

---

[1] Credit is due to the late G. F. Knapp for having drawn attention to Fourier's contributions: cf. *Theorie des Bevölkerungs-Wechsels*, Braunschweig, 1874, pp. 78 *sq.*

[2] Günther, *Geschichte der Deutschen Statistik. Die Statistik in Deutschland*, I, 1911, pp. 39 *sq.*

[3] *Abriss der neuesten Statistik des preuszischen Staates*, 1804, 2nd ed., 1805.

[4] *Betrachtungen über den National-Reichthum des preuszischen Staates*, 1805.

Physiocracy, to calculate the net income of the population. His investigations aroused the attention of the Prussian King (Friedrich Wilhelm III), who took a fancy to the establishment of a statistical bureau, with Krug at the head of it. This plan was realised in 1805, but in this shape the bureau had only a very short existence. A report was published in 1806, with several tables on population, agriculture, commerce, etc., but already in the same year—after the Battle of Jena—the work of the bureau was suspended. Publication of statistics on various important matters (such as income, public debts, etc.) was prohibited, and the material which the bureau possessed was removed to Copenhagen. The bureau was not re-established till 1810. This time J. G. Hoffmann (1765–1847) was appointed director, whereas Krug only got a secondary position. Undoubtedly Hoffmann was a more prominent man than Krug, so that the latter could not complain of this arrangement as an injustice.

E. Engel, who many years after held the same position as director of the Prussian statistical bureau, expresses his admiration for Hoffmann in very high terms.[1] If the present generation of statisticians is. to judge the opinion of Engel, the administrative qualities of Hoffmann will perhaps receive highest mention. The organisation of public statistics seems to have been a success. On the whole, *German* administration seems to have overcome the difficulties which naturally arose in those days with regard to official statistics, and *Prussia* was in this respect one of the leading countries. The distribution of the population according to age was not yet satisfactory (only three or four classes), and the causes of deaths were rather vaguely given. But at all events a foundation-stone had been laid on which the next generation could build. It must also be acknowledged that Hoffmann's various statistical reports give a fair description of the

---

[1] Engel, "Zur Geschichte des königlich preussischen statistischen Bureaus," *Zeitschrift des kgl. preussischen statistischen Bureaus,* Jahrgang 1, Berlin, 1860–1. He speaks of him as the hitherto "noch unerreichten Statistiker."

Monarchy.[1]   But in regard to finer theoretic investigations he was not particularly prominent.

Nor was Krug particularly clear on theoretic questions. A small report in manuscript is interesting in this respect. It was written in 1826 and a laconic remark is added by Hoffmann, a permission to print it.   Krug tries to find the numbers of inhabitants below 10 years of age at the close of the year 1825.   The problem is nearly the same as that which Halley solved for Breslau.   He assumes that $\frac{2}{3}$ of the infants who die in a calendar year under 12 months of age are born in the same year.   Under the supposition that emigration and immigration are trifling, he then finds the number of infants alive by means of the births and deaths.   For the next years of age he constructs the ages distribution of the deaths with the help of the figures for Berlin, which are more detailed than in any other part of the kingdom, and he further uses the same hypothesis that $\frac{2}{3}$ of the deaths pertain to one generation and $\frac{1}{3}$ to the preceding one.   Here, however, the calculation fails; when the age advances the numbers of deaths in each year of age approach to each other, and it will therefore be incorrect to use the proportion of $\frac{2}{3}$.

Other *German* States also made progress with regard to official statistics, even though these beginnings were rather small.[2]   Thus statistical topographical offices were established, in *Bavaria* 1808, in *Württemberg* 1820.   The efforts of Prussia to found a German Tariff-Union naturally led to an evolution of official statistics in several of the smaller *German States*,[3] particularly regular enumerations of the population, and as the Tariff-Union came into existence in 1833, triennial enumerations were prescribed for the whole territory of the Union.

---

[1] As, for instance, a small volume from 1818 (2nd ed., 1819): *Uebersicht der Bodenfläche und Bevölkerung des preuszischen Staates, aus den für das Jahr ämtlich eingezogenen Nachrichten.*   Later appeared *Beiträge zur Statistik des preuszischen Staates 1821 und Neueste Uebersicht nach den zur Ende des Jahres 1831 ämtlich aufgenommenen Verzeichnissen*, Berlin, 1833.

[2] Würzburger, "The History and Development of Official Statistics in the German Empire," *The History of Statistics*, 1918, pp. 334 *sq.*

[3] Meitzen, *Geschichte, Theorie und Technik der Statistik*, Berlin, 1886, p. 39.

An interesting combination of topography and statistics characterises the evolution in some of the *German States*. In *Württemberg* a long series of local descriptions began in 1824, the last volume (Number 64) appearing in 1886.[1] The volumes contain detailed topographical descriptions of history, fauna, customs and manners, etc., etc., but there is also fairly good information as to population and other statistical matters; the enumeration in 1822 had an age distribution into six classes for males, whereas females are only divided into two classes (above and under 14 years); later, in 1832, there was a distribution in ten groups of age. These descriptions are particularly interested in the increase or decrease of the population through wanderings. In addition to these Memminger published year-books from 1818 with contributions from several authors, on history, meteorology and statistics. Thus the year-book 1826 has contributions on Vital Statistics by Professor Schübler.[2] He has a curious calculation of the number of years in which a population is reproduced, viz. the length of time in which the number of deaths is equal to the population, whereas these again are replaced by births, a rather obscure definition. This number of years is found by dividing the yearly number of deaths in the population; if there is excess of deaths over births the number of births has to be used as denominator.

47. As mentioned in art. 31, *England* had a general census in March, 1801. There was, however, no central authority for the control of census operations.[3] The census was carried out in *England* by Overseers, acting

---

[1] The first one has the title: *Beschreibung des Oberamts Reutlingen. Herausgegeben aus Auftrag der Regierung von Professor Memminger, Mitglied des Königl. Statistisch-Topographischen Bureau,* Tübingen, 1824.

[2] Schübler, *Ueber die Gesetze der Bevölkerung und Sterblichkeit oder die Verhältnisse des physischen Lebens der Einwohner Württembergs. Württembergische Jahrbücher für Vaterländische Geschichte, Geographie, Statistik und Topographie,* 1826.

[3] *Official Vital Statistics of England and Wales,* League of Nations' Statistical Handbooks, Geneva, 1925, pp. 17 *sq.*; Baines, "The History and Development of Statistics in Great Britain and Ireland," *The History of Statistics,* 1918, pp. 367 *sq.*

under Justices of Peace and High Constables, in *Scotland* by Village Schoolmasters and others, whereas *Ireland* was excluded from the Bill.

The questions which had to be answered were very few, only (for each parish) the numbers of houses, families and persons of each sex; further, the distribution according to employment in three groups (agriculture, trade, and other callings).

Few changes were made by the succeeding censuses of 1811 and 1821.   In the latter a curious attempt was made at finding the distribution according to age (in certain age-groups).   In point of fact, the return was quite voluntary, it was only made "if not inconvenient to the parties."   John Rickman (1771–1840), who has a good name in the history of statistics of England, made reports of all these enumerations, as also of the following census of 1831.

How good the soil in England and Scotland was for statistical reports was shown by John Sinclair, who in a comprehensive work of twenty-one volumes gave a statistical account of Scotland, with detailed descriptions of each parish, supplied by the Scotch clergy.[1]   These descriptions are of diverse value, also the contributors do not follow the same schedule as to population statistics. Sinclair made no attempt at a general report, but in the "contents" of the various volumes there is a summary of the population in the parishes concerned.

In the *United States of America*, as we have seen in art. 32, a general census was prescribed in the constitution.   The census of 1800, as well as that of 1810, had five age-groups, for each sex separately, of the free population, whereas the number of slaves was returned without any distinction as to sex or age.   At the census of 1810 the marshals were instructed to take "an account of the several manufacturing establishments and manufactures within their several districts."[2]   A digest of this

---

[1] *The Statistical Account of Scotland drawn up from the Communications of the Ministers of the different Parishes,* Edinburgh, 1791–9.

[2] John Cummings, l.c., pp. 671–3.

census was taken by Tench Coxe. At the census of 1820 slaves and free coloured persons were returned by sex and age, and further items were added, for instance, the number of foreigners. The following census, of 1830, for the first time used uniform printed schedules. The white population was distributed in quinquennial ages, whereas the free coloured persons and the slaves were returned by six age-classes. Further, there were returns of blind, deaf, and dumb persons. There was thus decidedly a progress, but still all these enumerations were more or less alike. In 1840 the American Census first aims at becoming an important, complicated instrument for the gathering of information, exhibiting "a full view of the pursuits, industry, education, and resources of the Country," in response to a message in 1838 from the President. Here, as in Europe, the thirties are a turning-point.

As there was no permanent statistical institution it was left to private persons to make use of statistical observations at hand, not only concerning vital statistics, but in all branches of social and economic life. Neither in Great Britain nor in the United States were private individuals hindered from publishing statistical facts, and the American literature shows interesting contributions of this kind. Here an extensive compilation by Adam Seybert (1818) can be quoted.[1] As the author remarks, it is built up on documents which were presented annually to Congress (more than 120 volumes). There are very detailed statistics of imports (quantity and value) as well as of exports, further tonnage and navigation, tonnage employed in fisheries, finance of the United States, etc., etc. Whereas immigration is rather easy to control, the weak point seems to have been mortality- and birth-statistics. He gives a table concerning expectation of life in the City of Philadelphia, probably only based on deaths.

[1] Ad. Seybert, *Statistical Annals embracing Views of the Population, Commerce, Navigation, Fisheries, Public Lands, Post-Office Establishment, Revenues, Mint, Military and Naval Establishments, Expenditures, Public Debt and Sinking Fund, of the United States of America,* founded on Official Documents, Philadelphia, 1818.

48. In *Scandinavian* countries progress can also be recorded. Among these Sweden had the lead as before. After some years' failing activity of the Tabulating Commission, as mentioned in art. 32, the astronomer H. Nicander had been appointed as its secretary (1790), and he took up his task with great energy, writing several reports in the *Annals* of the Academy, 1799–1814. In one of these papers he improved Wargentin's method; instead of using the last number of inhabitants in a triennium, as Wargentin had done, in calculating a table of mortality, he chose an average, thus bringing the rates of mortality nearer to the truth. In 1802 some improvements in the schedules were made. Thus the marriage-statistics were altered, the infant mortality of illegitimate children was observed, and a sharp distinction was made between urban and rural population. Important as such improvements were, it was still more valuable that the annual tables hereafter were to contain information about the *sowing* and *harvest* of the most important cereals and vegetables, and the quinquennial tables to give approximate numbers of horses, cattle and sheep, as well as the approximate area under cultivation. The clergy, however, complained energetically of the new duties which these schedules laid on their shoulders, and at last they succeeded in so far that the agricultural part of the schedules disappeared from 1821. For many years hereafter agricultural statistics depended on reports from the provincial governments. Another step backwards was taken in 1830, the clergy being released from stating the *causes of death* which had been given since 1750. Only violent deaths had hereafter to be recorded. On the other hand, statistics of *still-births* were improved, from 1831. Beginning with the year 1804, the numbers of persons *vaccinated* were recorded.[1]

In *Denmark* conditions were less favourable than in Sweden. The Tabulating Office, established in 1797,

---

[1] Cf. Arosenius, *Bidrag till det svenska Tabellverkets Historia*, 1928, and "The History and Organization of Swedish Official Statistics," *The History of Statistics*, 1918, pp. 547 ff.

worked with very little energy. The Office was to revise and analyse the public accounts, and the census of 1801 having been taken, it became its chief duty to compile lists of the population. But the bureau was abolished in 1819, without having yet concluded its work with regard to the census. In 1834 a new era first began by the creation of the "Tabulating Commission," which will be mentioned below.[1]

While still united with Denmark, *Norway* had an enumeration in 1801. After its separation from Denmark in 1814, a new census was taken in the following year, and again ten years later, in 1825. Moreover, tabular surveys were prepared of import, export and shipping, and in 1829 a report was published on the economic condition in the various provinces, with reference to agriculture, forestry, fishing, etc. There was no separate statistical office till 1837; before that time the statistical work had been placed under the Department of Finance, Commerce and Customs.[2]

What has been said here may be sufficient to show the state of official statistics about 1830. Of course there were several other attempts at organising a statistical service. Thus in the *Netherlands* a summary census was taken in 1795; a royal decree of 1828 introduced a regular decennial census (the first one taken 1829), and from 1812 registers of births, deaths and marriages were kept in all communes. In 1825 a royal decree prescribed regular statistics of exports and imports, and from 1826 a statistical year-book appeared regularly.

In 1826 a statistical bureau was established under a committee of three members, with E. Smits as secretary. But on account of the separation of *Belgium* from *Holland* the bureau soon ceased to work. Smits left for Belgium, and the reorganisation of official statistics in *Holland* was postponed till 1848. In *Belgium* a general statistical

---

[1] Ad. Jensen, "The History and Development of Statistics in Denmark," *The History of Statistics,* 1918, pp. 201 *sq.*

[2] A. N. Kiaer, "The History and Development of Statistics in Norway," *The History of Statistics,* 1918, pp. 447 *sq.*

bureau was founded in 1831. Smits and Quetelet were charged with publishing a report on the census of 1829.[1] The latter had by this time already won a name as a statistician.[2]

---

[1] Cf. Julin, "The History and Development of Statistics in Belgium," *The History of Statistics*, 1918, pp. 125 *sq.*; Verrijn Stuart, "The History and Development of Statistics in Netherlands," *ibidem*, pp. 429 *sq.*, and Methorst, "Zur Geschichte der niederländischen Statistik," *Allg. statistisches Archiv*, 1903.

[2] By his *Mémoire sur les lois des naissances et de la mortalité à Bruxelles* (1825) and *Recherches sur la population, les naissances, les décès, les prisons, les dépôts de mendicité, etc., dans les royaumes des Pays-Bas* (1827).

CHAPTER XII

THEORY OF STATISTICS, 1800 TO 1830

49. Political Economy had in the eighteenth century grown to be a science. Quite naturally it might have been expected that this science would co-operate closely with statistics and thus promote the growth of the latter. This might particularly be expected when political economy turned to the problem of population, and a discussion on this subject was opened on the threshold of the nineteenth century by R. Malthus' *Essay on the Principle of Population*,[1] a discussion which for generations was one of its most characteristic features. This was, however, unfortunately not the case.

Malthus himself made no original contributions to statistics; his knowledge was only second-hand, and he refers largely to Price, Süssmilch and others. It is even striking how weak his statistical conclusions are. Thus one of his chief arguments refers to the growth of population in *North America*. He remarks (*Essay*, 1798, pp. 105–6) that the original number of persons who settled in the four provinces of New England in 1643 was 21,200, and he adds: "Afterwards it is supposed that more left than went to them. In the year 1760 they were increased to half a million. They had therefore all along doubled their own number in 15 years."

It is not difficult to see the weakness of this argument. It is evidently not enough to know that emigration in the long run surpassed immigration, the question being at which time did these movements take place. 21,200

---

[1] *An Essay on the Principle of Population*, 1798. It was reprinted in 1926, with notes by James Bonar. The second edition (followed by several others) appeared in 1803, very much enlarged, in two volumes.

immigrants settling shortly after 1643 would contribute nearly as much to the growth of population as the first colonists, whereas an afflux of the same size in 1760 would be relatively insignificant. It will be easy to find plausible suppositions for the movements of population which will lead to results quite opposite to those which Malthus assumes.[1] This does not make void the chief conclusion that a human population may gradually reach an enormous size, but it must be admitted that it makes a deeper impression if a couple of men after 250 years have 2,000 descendants (doubling in 25 years), than if the rate of increase is, for instance, 1 per cent. yearly, and the number after 250 years consequently is only 24.

Nor was Malthus fortunate in his choice of other evidence. He quotes Petty and his fantastic hypothesis of a doubling in 10 years, and Euler's theoretic calculation as to how long the doubling period is under certain suppositions, one of these giving as result 12–13 years.[2]

Malthus was well aware that there are great differences in the numbers in vital statistics, mortality for instance varying from 1 in 20 to 1 in 10, it being consequently useless to obtain a general measure of mortality for all countries taken together (l.c., I, p. 387). But on the other hand he adheres to the idea that duration of life is constant: "With regard to the duration of human life, there does not appear to have existed from the earliest ages of the world to the present moment the smallest

---

[1] A population of 21,200 growing regularly at a rate of $2\frac{3}{4}$ per cent. yearly will in the course of 117 years reach the size of 500,000. A natural increase through the surplus of births over deaths will hardly in the long run attain such a height as that. Supposing 2 per cent., which is still an uncommonly high rate, the 21,200 first settled will have increased to about 215,000 after 117 years. If, again, 35,000 immigrate about 5 years after the first colonisation, this will amount to about 322,000 in 1760; if at this moment 37,000 leave the colonies, a little more have emigrated than settled in 1643. Of course, such calculations can be varied infinitely many times, but they will at all events show the weakness of an argument, which does not take the moment of the emigration or immigration into consideration.

[2] 3rd ed., 1806, I, p. 7.

permanent symptom or indication of increasing prolongation" (l.c., II, p. 89).

It was at that time hardly possible to procure statistical observations as an argument for or against the alleged constancy of the duration of life. Malthus was confined to an abstract reasoning: "Nature," he says, "in the attainment of her great purposes, seems always to seize upon the weakest part. If this part be made strong by human skill, she seizes upon the next weakest part, and so on in succession." He adds that "the removal of any of the particular causes of mortality can have no further effect upon population than the means of subsistence will allow" (l.c., II, pp. 363–4). He quotes Heberden, who had found it impossible to explain changes in the frequency of various causes of death.[1]

For Malthus, as we have seen, the explanation was not far to seek. Just at this epoch the greatest hygienic measure in the nineteenth century was going to be taken, i.e. general vaccination against small-pox, and the effect was striking. Several authors can be quoted who at once acknowledged this and expected a considerable prolongation of the duration of human life. Duvillard is an example, and in actuarial literature others may be quoted. But Malthus and his followers only expected an increase of other causes of death as compensation for decrease in mortality from small-pox. Thus he says:

"The small-pox is certainly one of the channels, and a very broad one, which nature has opened for the last thousand years, to keep down the population to the level of the means of subsistence, but had this been closed, others would have become wider or new ones would have been formed. For my own part I feel not the slightest doubt, that if the introduction of the cow-pox should extirpate the small-pox, and yet the number of marriages continue the same, we shall find a very perceptible difference in the increased mortality of some other diseases" (l.c., II, pp. 366–7).

These quotations show Malthus' views. To a great

---

[1] Heberden, "Observations on the Increase and Decrease of Different Diseases and particularly of the Plague," 1801 (l.c., I, p. 465).

extent they were shared by economists three or four
generations ago.   J. B. Say (1767–1832), one of the
leading economists of his time, expounds just the same
theory.   He quotes an English physician, Watt, who
found a decrease in the mortality from small-pox in
Glasgow, 1783–1813, and a corresponding increase of
mortality from other causes of death.   Watt concluded
from these facts that vaccination aggravated or even
caused other diseases, but Say finds it much easier to
say, that when Death has found one of the ordinary
doors closed, it will open others.[1]   Unfortunately there
is a lack of clearness in his discussion of this problem.
He gives a definition of the mean duration of life (*la vie
moyenne*) which most of his contemporary colleagues
would accept, as the average age at death of a large
number of deceased persons, whereas the probable
duration of life (*la probabilité de vie*) is found as the
average number of years which a number of persons of
a given age will expect to live.   The former is—if
correctly calculated—the expectations of life at birth,
the latter the expectations at some other age.

The problem of mortality was, so to speak, reduced
to a question of arithmetic by Malthus and J. B. Say.
It did not occur to these authors that the problem is
much more complicated than a *regula-de-tri*.   Elasticity
in the consumption of food will allow great variations in
the movements of population without requiring an
increase of some causes of deaths as a compensation for
a decrease of others.   On the whole, statisticians and
economists in those days were rather naïvely inclined to
adopt quantitative theorems.   It is sufficient to recall
the wages-fund theory, not to speak of the theory of the
correspondence between prices and money in circulation,
or the supposition that each working-hour of a day in
a workshop gives the same output (one of the main
arguments in the ten-hour discussion).   A dogmatism
of this kind was of course not favourable to unbiassed

---

[1] J. B. Say, *Cours complet d'économie politique pratique*, IV, Paris, 1829,
pp. 379 *sq.*

statistical investigations.   As long as such shallow quantitative theories prevailed in economy there was no stimulus to deeper discussions on these questions.

50. No wonder that we also meet quantitative theories outside of economy.   The *German* medical statistician Casper wrote a book in 1825 in which he denied that the introduction of vaccination would cause an increase of other diseases.[1]   But ten years later he expounds the theory that mortality chiefly depends on the number of births.   The birth-rate will constantly keep pace with death: marriages regulate death.[2]   The English author Th. Sadler, who criticised Malthus, has a similar theory: "The Prolificness of Human Beings otherwise similarly circumstanced, varies inversely as their numbers." [3]   His contributions will be discussed in a following chapter.

The *actuaries* of this period did not generally share the theory of the economists with regard to the stability of mortality, and—sometimes with reservation—they were generally aware of the variabilities in vital statistics. But it was quite natural that they were induced to try to find mathematical expressions for the relation between age and the rates of mortality, or as it was termed: for the *law of mortality*.   De Moivre's above-mentioned hypothesis is one instance, his formula being the simplest of them all.   A much more complicated formula was given by the *German* mathematician Lambert (1728–77).[4] It may be objected to this solution, that he did not succeed in fitting the formula at the same time to childhood and to other parts of human life.   Exactly the same objection can be made to a very simple formula proposed by Benjamin Gompertz (1779–1865).   Having already touched the problem in an investigation of 1820,[5] he

<hr>

[1] J. L. Casper, *Beiträge zur medicinischen Statistik*, Berlin, 1825, p. 192.

[2] J. L. Casper, *Die wahrscheinliche Lebensdauer des Menschen*, Berlin, 1835, pp. 191, 216.

[3] Th. Sadler, *The Law of Population*, 1830, II, p. 352.

[4] Westergaard, *Die Lehre von der Mortalität*, 2nd ed., p. 200.

[5] B. Gompertz, "A Sketch of an Analysis and Notation applicable to the Estimation of the value of Life Contingencies," *Philosophical Transactions of the Royal Society in London*, 1820.

took it up again some years later.[1] Not the least interest is due to his application of the continuous method. His statement is perhaps not quite clear, when he says (l.c., p. 518):

"If the average exhaustion of a man's power to avoid death were such that at the end of equally infinitely small intervals of time he lost equal portions of his remaining power to oppose destruction which he had at the commencement of those intervals, then at the age x his power to avoid death, or the intensity of his mortality might be denoted by aq^x, a and q being constant quantities."

Trying to interpret this definition, we may suppose that the force of mortality—as a measure of the destruction of human beings in the age concerned—is increasing geometrically, the equation being $\mu_x = a + b.c^x$, where μ_x is the force of mortality, and a, b and c constants.[2] There are here two co-existing causes, one of them independent of age (leading to death without a "previous disposition to deterioration"), the other one increasing with age (connected with an "increased inability to withstand destruction"). Leaving out of consideration the former, he arrives at a very simple formula: $\mu_x = b.c^x$, allowing a great simplification of all actuarial calculations, for one life, still more for two and more lives. It may be objected that the formula is not fit for lives below 20, or above 80, but in the practice of life-offices this is generally of no great consequence. Many years later (1860) Makeham took both groups of causes in view, and his formula allowing the same simplification has had a great influence in actuarial science. Few statisticians will nowadays look upon such formulæ as expressions of a physiological law, even if they have a due claim on attention as practical instruments of calculation.

---

[1] "On the Nature of the Function Expressive of the Law of Human Mortality, and on a New Mode of Determining the Value of Life Contingencies," *Phil. Trans.*, 1825.

[2] The number of surviving will then be $l_x = ks^x(g)^{cx}$, where k, s, g and c are constants.

Several other mathematicians entered into the problem about the same time, as Littrow (1832) and Moser (1839).[1] A formula by Edmonds (1832) was very much akin to that of Gompertz,[2] whereas Thomas Young gave a rather complicated solution.[3]

Less known are corresponding investigations with regard to *disease*. The problem had already been treated by Price under the preparation of a scheme for sickness insurance. There were not many positive observations available from friendly societies. Price felt justified in assuming that on an average one out of 48 members of friendly societies under 32 years of age would at any moment be disabled through sickness or accident (viz. 7·6 days yearly). In the following periods of life he assumed a gradual increase up to one out of 24 between 59 and 64 years of age. He founded his theory on the decreasing vital power according to the Northampton table, the probable duration of life being 28 years for a person aged 30, and 13 at the age of 60.[4] Edmonds (in the above-mentioned investigation) gave a similar formula based on the hypothesis that every case of death corresponded to two years of disease. But such formulas were soon made superfluous by detailed experience of friendly societies. On the initiative of Oliphant several Scotch friendly societies gave reports on age distribution and sickness of their members, the results being published in 1824.[5] This report is particularly interesting on account of its attempt at finding the effect of protracted diseases. It tries to calculate how many weeks of disease shall be allotted to the first 3 months of illness, to the next 3 months, to 6–9 months, and to the following time.

---

[1] Westergaard, l.c., p. 88 *sq.*

[2] *Life-Tables founded upon the Discovery of a universal law regulating the existence of every human being . . .*, London, 1832.

[3] "A Formula for Expressing the Decrement of Human Life," *Phil. Trans.*, 1826.

[4] Morgan's edition of *Reversionary Payments*, 1812, II, pp. 473 *sq.*

[5] *Report on Friendly or Benefit Societies exhibiting the Law of Sickness as deduced from Returns by Friendly Societies in different Parts of Scotland.* Drawn up by a Committee of the Highland Society of Scotland, Edinburgh, 1824.

In the period treated in this chapter *actuarial science* made good progress, particularly in England, and the technical processes were much improved, so that the invention of a "law of mortality" allowing a simplification of the calculations was felt less necessary. As mentioned in art. 36, Tetens invented the method of D and N columns, but his work remained unknown to English actuaries. In 1813 the astronomer Francis Baily (1774–1844) published a work (*The Doctrine of Life-Annuities and Assurances*) in which was inserted as appendix a paper which was read before the Royal Society in the preceding year.[1] Here he refers to laborious calculations by George Barrett, who in vain had tried to get them published. Barrett uses a similar method to Tetens. It was later with a small modification used by Griffith Davies in a work published in 1825.[2] In the following year Charles Babbage published a work in which he mentions Barrett's method which he has applied to the Carlisle table.[3]

As to statistics, the most interesting work of this period is Joshua Milne's above-mentioned (art. 33) work, in which the Carlisle experience is treated.[4] The basis of his calculations is the enumerations of the population, January, 1780, and December, 1787 (7,677 resp. 8,677 persons), and the lists of deaths, 1779–87 (nine years). As the mean number of persons exposed to death he takes the average of the two enumerations, though the census of 1780 was taken a year after the beginning of the period of observations, but he argues that the population in the beginning of 1779 was nearly the same as in 1780, on account of the large number of

---

[1] "On a New Method of Calculating the Value of Life Annuities."

[2] *Tables of Life Contingencies containing the Rate of Mortality among the Members of the Equitable Society*, London, 1825.

[3] *Comparative View of the various Institutions for the Assurance of Lives*, London, 1826, pp. 60 *sq.* and Table XI. Concerning the history of this interesting episode of actuarial science, confer L. Iversen, *Dødeligheden blandt Forsørgede*, København, 1910, pp. 49 *sq.*; H. Braun, *Geschichte der Lebensversicherung und der Lebensversicherungstechnik*, Nürnberg, 1925, pp. 236–40.

[4] J. Milne, *A Treatise on the Valuation of Annuities and Assurances on Lives and Survivorships*, London, 1815.

deaths in 1779.   Dividing the annual number of deaths by the number of living, plus half the number of deaths, he arrives at the probability of dying within a year, a formula which had a widespread application afterwards. The Carlisle table has one-year intervals, for the first year even shorter intervals; this was not the case with Heysham's material, which has five- or ten-year intervals, only with exception of the deaths below 5 years.   Milne was therefore obliged to adjust the observations.   This was done in a rather imperfect way by means of a graphic method; thus the rates of mortality in the three last quarters of the first year are increasing, in the years 46–50 they are decreasing, etc.   This, however, does not prevent the practical application of the table for assurance purpose.   But then the question arises whether the rates of mortality were not somewhat too high.   In fact, Milne makes investigations as to the decrease of mortality, particularly on account of the vaccination against small-pox, which, however, was chiefly dangerous to children under 5 years of age; moreover, other changes may go in the opposite direction.   Like J. B. Say (art. 49), Milne quotes Robert Watt (l.c., art. 776), and he finds it probable that "where the small-pox precedes the measles it produces such a change in the constitution as to render that disease milder."

Credit is due to Milne not only for having calculated the Carlisle table, but for the interesting investigations of his work on the whole.   Thus he criticised the results in statistical literature concerning the mortality according to conjugal condition.   Further, he collects observations on the influence of corn prices on marriages and births. He maintains, as a result of his investigations (art. 692), that any "material reduction in the price of wheat is almost always accompanied by an increase both of the marriages and conceptions, and by a decrease in the number of burials."   It may, however, be objected that the material for these investigations was rather imperfect.

For a student of the history of statistics Joshua Milne's

work is particularly useful on account of its details with regard to statistical literature.

Other contributions deserve note as testifying to the progress even in this dull period. Griffith Davies was mentioned above. In his *Tables of Life Contingencies* (1825) he compares the Northampton table with the experience of the Equitable Society. After corrections in accordance with this material the table served as a basis for the German life office in Gotha (*die Lebensversicherungsbank für Deutschland zu Gotha*) which began its life on New Year's Day, 1829, and had much influence on the evolution of life insurance on the Continent.

Another prominent work was a report by John Finlaison (1783–1860) on the experience concerning government life-annuities, which appeared in 1829.[1] The material embraces various tontines, thus King William's Tontine, which began 1693, the last person dying in 1783, three Irish Tontines (1773–78), the Great English Tontine (1789), also Life Annuities of the Sinking Fund (1808).

The institution of the government annuities as a means of obtaining income for the State proved fatal, as the Northampton table, on which the tariff was based, had too high rates of mortality, the prices of the life-annuities being consequently too low. Finlaison's report on this subject having been printed in 1824, he was directed to make further investigations, and the above-mentioned report was the result. The method which he used in treating the observations (including a simple mechanical adjustment) is explained very clearly, the result being various life tables which in the report are compared with other existing tables. It is interesting to notice the difference among the older and younger experience, mortality in the tontine of 1693 being, for instance, higher than in the tontine of 1789. In addition, he made an investigation in regard to the sickness in a

---

[1] *Report of John Finlaison, Actuary of the National Debt, on the Evidence and Elementary Facts on which the Tables of Life Annuities are founded.* Ordered, by the House of Commons, to be printed, 31 March, 1829.

friendly society in London, which he compares to the above-mentioned Scotch statistics.

John Finlaison makes interesting remarks on various causes which influence mortality. He is very cautious with regard to conclusions from his material, pointing to the fact that his observations are "restricted to the higher and more affluent orders of society," and he maintains that the "rate of mortality affecting exclusively the industrious and the labouring and indigent classes is altogether unknown." But he remarks that "it may perhaps be concluded from the Carlisle table, that there is very little, if any, heavier rate of mortality affecting the lower than that affecting the affluent classes, notwithstanding the many advantages that these can command." As to the influence of vaccination against small-pox he has no material to judge of the movements of infant mortality, but he adds that it is not "in his power to believe that the small-pox in that period was so fatal a scourge as it is supposed, unless indeed its whole severity fell on infants under three years old and on the children of the poorer classes exclusively." In harmony with the opinion of J. B. Say and others he asks in addition "whether, after all, the small-pox itself might not have been the development of some evil principle in a particular form, which, opposed and put down in that form, might not still be in a state of activity in some other."

Finlaison's report thus points to important unsettled questions, but actuarial and statistical science had at all events reached a standpoint which guaranteed earnest scientific investigations of these problems and of several other questions which gradually appeared in connection with life-insurance.

CHAPTER XIII

THE ERA OF ENTHUSIASM
1830 TO 1849

51. In the history of statistics the two decades 1830–
49 can justly claim the character of enthusiasm. Of
course no progress would be possible in any scientific
branch without this quality, nor was any period of the
history of statistics since its birth in the seventeenth
century deprived of it; it will suffice only to quote
Süssmilch's *Göttliche Ordnung*. But enthusiasm is particularly evident in the two decades concerned. During
this period statistics attracted public interest to an
unusual degree. Official statistical institutions were
founded or re-established in several countries, and
numerous statistical societies sprang up and worked
in co-operation with these institutions. Statistical journals were started. Highly talented authors, as, for instance, A. Quetelet, made the statistical results accessible in lucidly written books, in which the apparently
dry observations were interpreted in an attractive manner.
Of course this was not without drawbacks: there was a
great temptation to dilettantism, and many statistical
publications were superficial and full of hastily acquired
results. No wonder that a reaction took place later
on, with a deeply rooted suspicion against the statisticians. It proved necessary to pull down several badly
constructed buildings and to replace them by more solid
structures. But statistics profited from this passing
popularity, inasmuch as large fields came under tillage,
and the mass of statistical observations got an enormous
increase.

A few facts will throw light on the evolution in this period.[1]

The most conspicuous progress was made in *England*. Here, in 1833, a statistical department was added to the Board of Trade. Much statistical material had been collected beforehand, but was generally of very little use; the greater part of it remained in the Offices without any movement towards its utilisation. Lord Auckland, who was minister 1830–4, wished to have a digest for the Board of Trade of the information contained in the Parliamentary reports and papers, and engaged G. R. Porter (1792–1852) for the task; soon after the Statistical Department was permanently established under his supervision. It was undoubtedly very fortunate that Porter became the head of this institution. The "incoherent mass of periodical tables then prepared was for the first time reduced to orderly and comprehensive returns, accompanied by lucid explanations of the meaning and limitations of the figures."[2]

Not less important was the establishment in 1837 of *civil registration* of vital statistics in England (later extended to Scotland and Ireland). Registrars were appointed throughout the Kingdom for the registration of marriages, births and deaths, with a Registrar-General at the head of the institution. The regular organisation of vital statistics was thus made possible, especially as also from 1841 the decennial censuses were gradually by special Acts also placed under the Registrar-General. In 1839 a young physician, William Farr (1807–83), who had won a name as author on medical statistics and hygiene, was appointed as "Compiler of Abstracts," and later promoted to Superintendent of the Registrar-General's office. His services in this position were of very great value; they will be mentioned below.[3]

---

[1] Cf. the above quoted, *The History of Statistics*, 1918, pp. 368, 292, 334, 338, 127, 447, 551.

[2] Baines, *The History of Statistics*, p. 374.

[3] About his life and work confer *Vital Statistics: A Memorial Volume of Selections from the Reports and Writings of William Farr.* Edited by Noel A. Humphreys, London, 1885.

As to *France*, the progress of official judicial statistics was mentioned in a preceding chapter, regular annual statistical publications by the Ministry of Justice beginning in 1827. The great interest which was taken by *French* and *Belgian* statisticians in this subject is partially explained by the relatively rich sources which were thus opened. But still more important was the re-establishment in 1833 of the Bureau de la Statistique générale, which was suppressed in 1812. It was Thiers who, as Minister of Commerce, took this step, after having obtained the authorisation of the Chambers. The bureau had charge of several important subjects, such as population, finance, foreign trade and prices. But there was no decided centralisation, various branches being treated separately. Thus a statistical service was created in 1844 and the following years in the Ministry for Travaux publics (for instance, for railway statistics).

There was also considerable activity in *Germany* in this period. The Tariff Union, founded 1833, required regular censuses in all the States which joined the Union, as the income from the tariffs had to be distributed according to the number of inhabitants. There were therefore regular triennial enumerations till 1866. In 1846 and 1861 the census even included occupation and industry; statistics of shipping, mines and production of salt were introduced.

The statistical service in *Prussia* and *Württemberg* was mentioned above. In *Saxony* an officially supported statistical association was formed in 1831; in 1850 it was taken over by the State.

In *Holland*, as mentioned in art. 48, a statistical bureau was established in 1826; it was to have charge of the census of 1829. The separation of *Belgium* from *Holland*, however, put an end to the activity of the bureau, which was not re-established till 1848. On the other hand, there was great progress in *Belgium*. The provisional Government organised a statistical bureau in the Department of the Interior (1831). There was an official co-operation between Smits, who was appointed director of the bureau, and Quetelet, lasting till 1841, when Smits resigned. They were commissioned to publish the results of the Census of 1830 (or 1829). A Statistical Commission was organised in 1841, with the object of controlling the various branches of statistics. Quetelet became its president, a position which he held up to his death in 1874. Several official reports appeared under the auspices of this Commission. Its work centred particularly around the Census of 1846, which will be mentioned below.

As to the *Scandinavian* countries, *Sweden's* official statistics had, as we have seen, already been organised long before this period. Still, interesting progress took place, for instance, with regard to statistics of still-births and marriage statistics. More conspicuous were the steps taken in *Norway* and *Denmark*. In *Norway* a statistical bureau was organised in 1837, as a tabulating office in the Department of Finance (later, in 1846, transferred to the Department of the Interior). It proceeded at once to publish previous census results, and in 1839 it further published reports on Agriculture, Commerce and Shipping (1835), followed by several later publications. In *Denmark* a Tabulating Commission was created in 1834. It worked with considerable energy, and several able reports were published, for instance, concerning the censuses of 1834, 1840 and 1845. The first of these census reports also contained the main results of the Census of 1801. The commission disappeared in 1848, and in its stead, after some deliberations, a Statistical Bureau was founded in 1850.

52. All these newly formed or reorganised official institutions had a highly interesting supplement in the *statistical societies* which were founded in several countries and which co-operated more or less intimately with the official institutions.

The most striking instance of this co-operation is the *Saxonian* Statistical Society. The Central-Committee of the Society was favoured by the Ministry as well as by other authorities. On the other hand, a considerable number of private persons interested in statistics assisted the Society.[1] Among these private co-operators were numerous physicians; an attempt at collecting medical statistics for 1838 was made by the help of fifty-nine physicians.[2] Many clergymen complied with the request of the Society to transcribe from the Church Register lists of deaths which were to serve as a basis for mortality tables. One of the reports of the Society describes its regular work [3]:

"The Directory of the Society collects, arranges, and enters in journals, registers, and other books for this purpose, all accurate information which is of such nature as to be serviceable to the purposes of the *State*. The facts are afterwards methodically transferred to separate ledgers, each appropriated to an especial subject; and those of peculiar importance, which present information directly useful to the public service, are extracted and laid before the Ministers of the Government; while such as offer a more general utility receive publicity in the pages of the periodicals."

In this way manifold material was collected on topography and climate, population, marriages, births and deaths, medical statistics, agriculture, prices, market places, justice and crime, etc. In fact, from its very start the Society differed very little from a regular statistical bureau. The change in 1850, when the Society became a public institution of this kind, was by no means a radical step.

Other statistical societies were generally more independent. One of the best known is the *American Statistical Association*, founded in Boston, 1839. There were many societies of the kind, particularly in *Great Britain* and *Ireland*. The most prominent of these was the *Statistical Society of London*, founded in 1834. The pre-

---

[1] See the preface of *Mittheilungen des statistischen Vereins für das Königreich Sachsen*, Erste Lieferung, Leipzig, 1831.

[2] *Medicinische Statistik*, 1838 (Mittheilungen . . . Lieferung 12).

[3] *Journal of the Statistical Society of London*, I, 1839, pp. 110 *sq.*

ceding year a Statistical Society of Manchester saw light. Several others appeared in the thirties, in Ulster, in Liverpool, in Glasgow, in Leeds and in Bristol, etc. There was a tendency to create statistical sections in scientific societies. The Statistical Society of Ulster was formed in 1837 by some members of the Belfast Natural History Society, founded in 1821 with Natural History and Topography as its subjects. The Royal Cornwall Polytechnic Society offered prizes for statistical essays. It is particularly interesting to see the influence of the British Association for the Advancement of Science. At the Meeting of the Association at Cambridge in 1833, a Statistical Section was formed; the meeting in 1836 at Bristol gave rise to the Statistical Society there, so also in Liverpool, where a Statistical Society was started 1 January, 1838. Everybody seemed to have got statistics on the brain!

Many motives may have been leading to the foundation of these societies, but the most striking one seems to have been the interest *in social problems*. Students of Economy have been inclined to look upon the thirties as a barren period, during which social problems were treated with cool indifference. This is not true; on the contrary, many minds were deeply impressed by such questions. Naturally they presented themselves in other shapes than nowadays. At that time, for instance, popular schools were often looked upon as the most important remedy against social evils. The volumes of the *Journal of the Statistical Society of London* testify amply to the interest in the education of the poor. But the members of these societies did not confine themselves to the school problem. They wished to learn the whole truth with regard to the condition of the poorer classes of society.

Thus the above-mentioned Cornwall Polytechnic Society maintains that

"the value of statistical information can no longer be doubted. It stimulates benevolence, and gives aim and effect to the energies of the philanthropist; it furnishes the legislator with materials on

which to form remedial measures for social happiness. . . .
From information thus furnished it cannot be questioned that the
public attention has been fastened, with an intensity never before
given to the subject, upon the physical and moral degradation of the
poorer classes in the metropolis and many of our large towns."

In just the same spirit the Statistical Society of Glasgow
was established. It was the object "to collect, arrange
and publish, facts illustrative of the condition and prospects, with a view to the improvement, of mankind."

Several investigations were made by *paid agents*,
whose duty it was to examine the condition of the inhabitants in some district or other. These investigations may often appear imperfect, encumbered as the
schedules were by a chaos of questions. But a student
of the social history of England will in these reports
find extremely valuable information as to the physical
and moral condition in these districts, the state of education, schools, hygiene, etc.[1]

The Meetings of the British Association gave the
opportunity for lectures on this subject; undoubtedly
much knowledge with regard to the misery of the poorer
classes was spread in that way.

But the members of the Statistical Societies did not
confine themselves to these subjects. On the whole,
they were interested in a number of economic and statistical problems. Thus the London Society tried to collect
material for a statistical account of the various strikes
and combinations. One committee was formed for vital
statistics, another for criminal statistics. Sometimes there
was competition from outside. The committee on vital
statistics asked the life-offices for observations on mortality of insured persons, but found that they had taken
the initiative themselves.[2] And of course there were

---

[1] As a single instance, "The Inner Ward of St. George's Parish" (*Journ.
Stat. Soc. London*, VI, 1843). There were 5,945 inhabitants, 616 families read
newspapers, 823 did not read, 26 were unascertained. The report gives details
about confession, how many families attend worship, how many children
attend day schools or Sunday schools, how many privies are "decent" or "not
decent," etc.

[2] *Journ. Stat. Soc. London*, II, 1839.

many difficulties in getting good observations. Trying to find the correlation between crime and education the reports were confined to superficial tests, such as the number of persons who, when marrying, signed their names in the marriage registers with marks.[1]

Like the Saxonian Society, English statisticians wished to co-operate with the Government in order to procure more perfect statistical material. A Committee of the London Society was appointed in 1839 to report on the best mode of taking the Census of the United Kingdom in 1841; some of the recommendations were adopted, whereas other propositions, as, for instance, on the health of the population, were laid aside.[2] It may be, however, that Parliament was right in declining to put too many questions. A few years before the Board of Trade made an unsuccessful attempt at a statistical report on the state of agriculture in Bedfordshire. An appeal was made to the resident clergymen, but only one out of five made returns, so the experiment had to be abandoned.[3]

On the other hand, the questions here seem to have been more complicated and puzzling than the proposed questions in the Census schedules, as, for instance: "What is the quantity and description of cheese made in the year?" The statisticians of those days had not sufficient experience to know what questions to put and how to formulate them. In so far this experiment is interesting as here, just as in *Saxony*, private persons were invited to assist in collecting statistical material.

But in spite of all difficulties, official statistics became more detailed and probably also more trustworthy in the period concerned. By the English Census of 1831 the census of ages was not enacted, only males above 20 were returned, distributed according to occupation.

---

[1] See, for instance, F. G. P. Neison, "Statistics of Crime in England and Wales for the years 1834–44" (*Journ. Stat. Soc. London*, XI, 1848).

[2] G. R. Porter, "An Examination of some Facts obtained at the recent Enumeration of the Inhabitants of Great Britain" (*Journ. Stat. Soc. London*, IV, 1841).

[3] Cf. a letter from G. R. Porter (l.c., Vol. I, 1839).

In 1841, however, the schedules demanded individual information regarding age, sex and occupation, and there was a distinction between foreigners and natives and between persons born in the districts and outside. It was a very important step forward, that whereas hitherto the Overseers were responsible for the entries, it now became the duty of the householders to fill up the Census schedules. This system has continued through all the following enumerations, with only slight improvements and additions.[1] In the above-quoted article on the Census G. R. Porter maintains that there was "a more correct account of our numbers than was attained by any former enumeration." The registration of *births* was not compulsory till 1874, so that an unknown fraction of the births, perhaps 5 per cent., was not recorded. But only a few *deaths* escaped registration after the establishment of the system of 1837. The registers had columns for particulars regarding sex, age and profession, as well as the cause of death. At first the causes of death may have been imperfectly stated, but from the year 1845 only certificates by men authentically qualified were recognised, and the material was therefore undoubtedly more accurate. Important improvements in the nomenclature were due to William Farr. Thus England and Wales came into possession of interesting observations on mortality, and English Vital Statistics justly enjoyed great reputation.

Less striking was the progress of vital statistics in *France*. The census of 1831 was probably not very reliable. There was no distribution according to age.[2] In 1836, however, a great step forward was made, the schedule now designating the inhabitants by family and by household, but it was not till 1876 that individual reports came into use. There was therefore in official

---

[1] Greenwood and Granville Edge, *Official Vital Statistics of England and Wales* (League of Nations' Statistical Handbooks, 3, Genève, 1925).

[2] Males are distributed among "Enfants et non mariés," "mariés," "veufs," "militaires"; females have only three classes: "enfants et non mariées," "mariées," "veuves" (*Documents statistiques sur la France*, publiés par le Ministre du Commerce, 1835).

statistics no basis for reliable mortality-tables.   More valuable are the publications on economic statistics, on import and export (official values), shipping, and, later, agriculture (according to circulars to the *préfets* of July, 1836).   The latter report, which appeared 1840–1, contains the results of inquiries in 37,300 communes, as to live stock, value and quantity of corn produced per hectare, seed, etc.

In *Belgium*, as mentioned above, Smits and Quetelet developed great activity.   The Census on New Year's Day, 1830, was treated by them in a remarkable report.[1] Progress continued under the auspices of the Central Commission.   The chief event was the census of October, 1846, embracing population as well as agriculture and industry.   There were details on houses, fire insurance, instruction, religion, language, age and civil status, etc.   In agriculture, the domestic animals were counted, there were observations concerning area under cultivation, crops, wages of day labourers, etc.   The industrial census had a detailed classification of professions; the report contains the number of working men, their wages, the amount of horsepower of engines, number of looms and other utensils employed.   The whole census was generally looked upon as a very important step forward.

Even though progress in *Germany* in this period was conspicuous, it was not as radical as, for instance, in *England*.   The *Saxonian* Statistical Society was mentioned above.   Its first report, 1831, contains details concerning the population.   There are only three age classes, but in the fourth report of a census in 1832 there are already eleven age classes.   But when some years later Leonhardi proceeded to calculate a mortality-table for Saxony, he did not venture to use these observations, but constructed his table from deaths only.[2] Nor was the Census of 1832 in *Württemberg*, with ten

---

[1] *Recherches sur la reproduction et la mortalité de l'homme . . . .* 1832.

[2] *Mittheilungen des statistischen Vereins für das Königreich Sachsen,* Lieferung 13, 1833; 17, 1848.

classes of ages, utilised in this way. In this period economic statistics fared perhaps better than vital statistics. There was an industrial census in *Saxony*, 1846, and the Statistical Society made energetic efforts to get exhaustive agricultural statistics.[1]

Norway, as mentioned above, also made good progress. The Census report, 1835, had a fairly detailed distribution according to age (thirteen classes); the same may be said of *Denmark*. The two countries rivalled each other with regard to agricultural statistics.

As the result of this brief sketch, it may be said in all fairness that official statistics in these two decades made considerable steps forward, not only as to the quantity of observations but also as to their quality.

53. There was at that time in *France* no *journal* exactly corresponding to the *Journal of the Statistical Society of London*. But the *Annales d'Hygiène publique*, which was started in 1829, was from the very first open to statistical investigations of various kinds.[2] Some authors wrote about the influence of profession on the duration of life, others treated health in prisons or lunatic asylums, or anthropometry, etc. Later on there will be opportunity to judge the value of these contributions. Various diseases were discussed from a statistical point of view, such as the causes of tumours, and there were observations on fractures, dislocations, etc. Here a peculiar phase of the evolution of statistics may be mentioned, namely, the discussion in medical literature of what was called the *numerical method*. The most prominent figure here was the famous French physician P. Ch. A. Louis (1787–1872).[3] His method may be described as a detailed study of each one of

---

[1] l.c., Heft 9, 1838.

[2] This was also the case with the English medical journal, *The Lancet*.

[3] Cf. here P. Ch. A. Louis, *Recherches anatomiques, pathologiques et thérapeutiques sur la Phthisie*, Paris, 1825; *Recherches . . . sur la maladie connue sous les noms de Gastro-Entérite . . .*, Paris, 1829, and his polemical treatise, *Examen de l'examen de M. Broussais, relativement à la phthisie, et à l'affection typhoïde*, Paris, 1834.

a certain number of cases of the disease concerned. Having thus got an intimate knowledge of each individual case, he was naturally induced to deepen his insight by grouping the observations. In doing so, Louis is justly entitled to the name of father of medical statistics which has been bestowed upon him. It must be acknowledged that these statistical attempts are often very clumsy, and that his observations are generally so sparse that no safe conclusion can be drawn from them. Thus with regard to the connection between diarrhœa and typhus he has a table based upon 17 cases. Or he collects 123 cases of phthisis in La Charité, and finds that 66 out of these are females, who thus "seem to be more frequently attacked than males." Sometimes the numbers of observation are so small that he cannot fail to see that they are without value. Thus, after having given details of three observations on the duration of the disease, he exclaims: "What can be concluded from so small a number of facts?" [1]

Even if the observations were sufficiently numerous there were often other defects which made the results irrelevant. The collection of observations in a hospital will generally lead to disappointment. But sometimes there are ingenious glimpses, as, for instance, when he discusses the problem of heredity in phthisis and requires two sets of tables of mortality, one based on observations on persons with consumptive parents, the other one with non-consumptive parents. And it was undoubtedly not unimportant for William Farr's future as a statistician, that he attended the lectures of Louis during his two years' residence in Paris.

Sometimes the mistakes of the master are made more obvious by his disciples. A striking example is a treatise by J. Pelletan.[2] He expects that it will not be without interest to see the complete application of the numerical method to the study of an important

---

[1] *Recherches . . . sur . . . Gastro-Entérite*, I, p. 186.

[2] J. Pelletan, "Mémoire statistique sur la pleuro-pneumonie aigue" (*Bulletin de l'Académie Royale de Médecine*, 1837).

disease. He bases his investigation on 75 cases of pleuro-pneumonia. These observations are subdivided to the last extremity, for instance, according to professions, in 39 groups. One of his tables is formed on two cases only.

Louis' embryonical statistics were, however, not accepted at once. His contemporary Broussais (1772–1838) attacked him vigorously. Louis replied in an uncommonly sharp essay (quoted above). Broussais' fame was by that time waning, and his attacks were therefore less important, but the opposition against the numerical method found support amongst other medical men. This will be seen from an interesting discussion in 1835 and the following years, the cause of which was a treatise by Civiale on *l'affection calculeuse*, of which a committee appointed by the Academy of Science, with Double as reporter, gave their opinion.[1] The report maintains against the numerical method that each person has his own individuality; problems in medicine are always individual, the facts always presenting themselves for solution one by one; the treatment in each case depends on a happy instinct supported by numerous comparisons and guided by experience. Double was evidently right in maintaining that each case has its individual character, but he failed to see that this does not exclude the application of the theorems of the calculus of probability, even on a limited statistical material, if it can be proved that these observations, in spite of individual differences, follow the law of error; if this is the case, the physician in considering all the individual circumstances will have a guide in choosing between various cures.

54. Medical statistics had to master two difficulties. One was to procure *unbiassed* observations; this will be discussed below. The other one was to decide whether the number of observations was *large enough*. We have seen how mathematicians like J. Bernoulli, de Moivre

---

[1] *Comptes Rendus Hebdomadaires des Séances de l'Académie des Sciences,* Paris, 1835.

and Laplace contributed to the solution of this problem. Several mathematicians of the day approached the question of the precision of statistical results, showing that this precision increased in the same proportion as the square root of the number of observations. Various authors may be quoted here. Thus the English mathematician Augustus de Morgan published a lucidly written essay on the calculus of probabilities.[1] Quetelet discussed the problem in his *Recherches statistiques* (1844) and in his *Lettres sur la théorie des probabilités* (1846), but strangely did not apply the theory much in practice. The well-known mathematician Cournot in his *Exposition de la théorie des chances et des probabilités* (1843) enters somewhat more into statistical problems, as, for instance, with regard to the balance between the two sexes. Still, there was no bridge between the mathematicians and the statisticians who entered into the question of the precision of the results, without deeper mathematical training. Nor did Poisson (1781–1840) succeed in this direction with his far-reaching work on the calculus of probabilities.[2] Curiously enough he was a member of the committee which reported on Civiale's treatise. He was chiefly interested in the results of judicial statistics, but his investigations also apply to medical statistics. His book is a remarkable extension of the calculus of probabilities to the problem of finding the degree of precision where there are two or more values, each with its own special circumstances; he makes it possible by a very simple formula to calculate the mean error of a difference between two observed frequencies. If, then, two treatments of a disease give different results, it is easy to ascertain whether we are justified in concluding from the observations in question that one of the cures is preferable. It may be inserted here that another important contribution to the calculus of probabilities was published a few years

---

[1] Augustus de Morgan, *An Essay on Probabilities and on their Applications to Life Contingencies and Insurance Offices*, London, 1838.

[2] Poisson, *Recherches sur la probabilité des jugements*, Paris, 1837.

later by the French astronomer Bravais (1811–63).[1]
His discussion of the correlation of two or three
variables was in harmony with the theory which many
years later was developed by English mathematicians,
on the correlation between two or more groups of
facts, as, for instance, age at marriage of bride and
bridegroom, colour of hair and eyes of father and
son, etc. But at present this treatise hardly seems
to have been noticed. In this respect Poisson was
more fortunate, for a few years after the appearance of his work, Gavarret, an enthusiastic pupil of
Poisson, published a book on medical statistics, in
which he popularised Poisson's results.[2] As a limit,
he chooses the mean error multiplied by $2\sqrt{2}$. If the
difference between two frequencies is above this limit
he considers it as proved that particular causes have
been active. Thus if there are two groups of cases,
500 in each, and if among them 100 and 150 respectively
were cured, whereas 400 and 350 respectively died,
then Gavarret's limit will be 0·077, the mean error of
the difference (0·30 − 0·20 = 0·10) being 0·027. As
the difference exceeds the limit we may safely conclude
that the former cure is less promising than the latter.
It will only require a few steps more to give this method
a very important applicability to medical statistics, on
one side, of course, an examination of the facts in order
to ascertain if they follow the law of error, on the other
hand, an extension to other limits. Gavarret, however,
does not proceed so far.

Unfortunately his contemporaries took very little
notice of his book. Ten years later the English physician William A. Guy—who was a very energetic contributor to the *Journal of the Statistical Society of London*
—wrote an article on the problem.[3] He arrives at the

---

[1] Bravais, *Analyse mathématique sur les probabilités des erreurs de situation d'un point* (Mémoires de l'Institut de France, Séances mathématiques et physiques, IX, 1846).

[2] Gavarret, *Principes généraux de statistique médicale*, Paris, 1840.

[3] W. A. Guy, "On the relative Value of Averages derived from different numbers of Observations" (*Journ. Stat. Soc. London*, XIII, 1850).

negative conclusion, that "the formulæ of the mathematician have a very limited application to the results of observation; and that if incautiously applied, they may lead to very great errors." It appears, however, from the article that he did not really grasp the meaning of Gavarret's formulæ. He is on safer ground when grouping observations of average ages at death and finding that the difference between the greatest and the least average derived from successive groups of facts diminishes rapidly as the number of facts increases. Guy had, however, not sufficient mathematical training for deeper investigations of this kind, nor did his contemporaries take much interest in the problem.

Of course, statisticians of those days could not help pondering over the problem of precision of statistical results. Sometimes the practical rule is recommended to divide the material in groups and compare the results derived for each of them. Thus W. A. Guy after having maintained that no general rule can be laid down as to the number of observations which may be sufficient in any particular case to avoid error or to determine a real average, adds that

"where our means or instruments of observation are imperfect, or the things observed differ widely in numerical value, a large number of observations is necessary, in other cases a smaller number can suffice. Perhaps the best rule which can be given for ascertaining whether the observations which we have collected are sufficiently numerous to yield a true average, is to divide the whole number of observations into groups of equal size and compare them the one with the other: if the average value of each group is the same, we may safely conclude that we have arrived at the true mean; if not, we must increase the number of our observations, and the size of our groups, till the desired equality is obtained. . . . But if it is important to derive our averages from a large collection of observations, it is not less essential to group together these observations only which bear a close resemblance to each other." [1]

---

[1] "On the Value of the Numerical Method as applied to Science, but especially to Physiology and Medicine" (*Journ. Stat. Soc. London*, II, 1839, p. 34).

It is not difficult to prove the defects of such practical rules; at all events they show that the problem was alive. Frequently the statistical results are criticised on account of the limited number of observations. We meet, for instance, this critical mind in a review of a report by Renaudin on the statistics of lunatics.[1] Unfortunately the reviewer, in maintaining that the number of observations is too small, does not explain how it can be testified whether the material is large enough. Often it was left to the tact and experience of the author, whether he would claim the reliability of his results. Sometimes there is a sort of competition, the report in question boasting of the large number of observations compared with the material at disposal by previous investigations. Or an arbitrary rule was introduced, for instance, when Villermé in an article on the health in some textile factories[2] calculates the probable duration of life in a trade only if more than 100 deaths have been observed.

In some cases the author himself makes excuses for the small number upon which the investigations are based. In an investigation of the height of men in France the author measures 27 persons aged 20–25 and 21 aged 25–30. He finds 1,679 and 1,697 mm. respectively. He agrees that the numbers of observations are small, but he adds that his results may nevertheless be found interesting.[3] Behind this excuse there may be hidden a wish to heap together homogeneous observations till at last a sufficient material has been provided.

But in most cases the statistical authors do not seem to be nervous on account of the smallness of their material, and the statistical literature of that period testifies to a remarkable lack of critical sense. Thus Marc d'Espine distributes 14 cases of death from scarlet

---

[1] *Annales d'Hygiène publique*, 27, 1842.

[2] Villermé, "Sur la santé des ouvriers employés dans les fabriques de soie, de coton et de laine" (*Annales d'Hygiène publique*, 21, 1839).

[3] F. Lélut, "Essai d'une détermination ethnologique de la taille moyenne de l'homme en France" (*Annales d'Hygiène publique*, 31, 1844).

fever according to months.[1] Guy deals with the influence of employment on health, and among other trades he mentions locksmiths and bellhangers, with 4 out-patients in King's College Hospital.[2] When measuring the duration of life of sovereigns he operates with observations on 9 Scotch kings.[3] Another example is found in an investigation by Benjamin Phillips on mortality in the case of amputation.[4] Altogether 640 cases of amputation of an arm or a leg are considered, with 150 deaths, in four different countries. *Great Britain* has a superiority over *U.S.A.* with a mortality of 0·228 and 0·253 respectively (53 and 24 deaths respectively). A modern statistician will at once calculate the mean error of the difference 0·025. The result being 0·05, he will lay the material aside, as it gives no proof of the alleged superiority. Something similar may be said of the investigations concerning the balance of the two sexes, by Hofacker and Sadler, which will be mentioned below; not to speak of a curious book by R. R. Madden[5] : the author collects facts about 20 deaths in each of 12 groups, drawing bold conclusions as to the average age at death in each of these groups.

55. Even if the problem of the law of error had not existed, other difficulties were—as remarked above—unavoidable in those days. In fact, many questions presented themselves for solution, which could not be solved by means of the material at hand, and the writers on medical statistics generally failed to see this. As a rule, they accepted the observations, imperfect and incomplete as they were, without criticising them,

[1] Marc d'Espine, "Essai statistique sur la mortalité du Canton de Genève pendant l'année 1838" (*Annales d'Hygiène publique*, 23, 1840).

[2] W. A. Guy, "Further Contributions to a knowledge of the Influence of Employment upon Health" (*Journ. Stat. Soc. London*, VI, 1843).

[3] W. A. Guy, "On the Duration of Life of Sovereigns" (*Journ. Stat. Soc. London*, X, 1847).

[4] *Journ. Stat. Soc. London*, I, 1839.

[5] R. R. Madden, *The Infirmities of Genius illustrated by referring the Anomalies in the Literary Character to the Habits and Constitutional Peculiarities of Men of Genius*, I–II, London, 1833.

drawing more or less bold conclusions from the material, and the result of all this naturally must be a great confusion. A problem which attracted many authors was to discover the influence of *occupation on health*. Nowadays a statistician, in order to solve this problem, would ask not only how many deaths or cases of disease, etc., took place in a certain period within the occupation concerned, but also how many persons have been exposed to risk. But material of this kind being as a rule not available, the investigation was mostly confined to a one-sided material, viz. the number of deaths or cases of disease in various classes of age. But owing to fluctuations in the working forces, the numbers exposed to risk varied considerably, and even if there was no difference in health the observation might show great apparent differences. The confusion became still more evident, when deaths from disease had to be sought in the registers of hospitals, so that only a section of the vital statistics of the country could be ascertained. The statistical literature of those days testifies amply to the great energy with which these subjects were treated, but unfortunately the positive results of these investigations were generally very meagre.

It would be superfluous to give a long list of authors who in this era of enthusiasm were engaged in studying this problem of health under various circumstances. A few examples will suffice.

The solution which quite naturally suggested itself was to prepare lists of deaths in different occupations or classes of society and compare the distribution according to age and the probable or mean duration of life. This was done in *Germany* by J. L. Casper in the above-quoted book on the probable duration of life.[1] He is well acquainted with the literature on the subject, but he does not discriminate between right and wrong methods of calculation of life-tables. He considers it sufficient to have lists of deaths only at his disposal; he maintains that the enumeration of a popu-

[1] J. L. Casper, *Die wahrscheinliche Lebensdauer des Menschen*, Berlin, 1835.

lation will involve too great difficulties. He collects 3,735 cases of death above the age of 23, in 10 professions, and he finds that the clergy have a high duration of life, the physician a small one. Just in the same way he compares the mortality in the German aristocracy (princes and counts) and that among paupers in Berlin who receive relief. In France Benoiston de Châteauneuf took up the latter problem.[1] It is not quite clear how he calculated the mortality of the aristocracy; as to the poor, he follows just the same line as Casper, distributing a number of deaths in the working classes in a district of Paris according to age. Had Casper proceeded to a comparison between his and Châteauneuf's death-rates he would perhaps have become a little suspicious on account of the lack of harmony between the two series of observations. In a later report Châteauneuf dealt with the mortality of men of science.[2] Though his own method is unchanged his paper has the great advantage that he has given all the details, so that it was possible later out of these observations to calculate correct rates of mortality.[3]

An author who contributed much to the confusion was H. C. Lombard.[4] In a paper dated 1835 he distributes 8,488 deaths in Geneva 1796–1834 according to profession, calculating the average age at death. Among these classes are students; no wonder that their average age is only 20. Lombard tries to solve a large number of questions, for instance, concerning dust inhalation, economic conditions, etc. Though his own investigations are badly founded, he displays some critical sense about other contributions. He criticises,

---

[1] Châteauneuf, "De la durée de la vie chez le riche et chez le pauvre (*Annales d'Hygiène publique*, 3, 1830).

[2] Châteauneuf, "De la durée de la vie chez les savans et les gens de lettre" (*Annales d'Hygiène publique*, 25, 1841).

[3] Westergaard, *Die Lehre von der Mortalität*, 2nd ed., 1901, p. 283.

[4] H. C. Lombard, "De l'influence des professions sur la durée de la vie" (*Annales d'Hygiène publique*, 14, 1835). See also "De l'influence des professions sur la phthisie pulmonaire" (*Annales d'Hygiène publique*, 14, 1834).

for instance, the above-quoted English author R. R. Madden, who discusses the duration of life of poets, musicians, philosophers, etc. There are 20 famous persons in each of 12 groups, and Lombard observes properly that these persons are selected, none of them being registered before they have made their name. Madden calculates the average age at death for these 12 groups, varying from 57 years (Poets) to 75 (Natural Philosophers). Laying this work aside, as more biographical than statistical, we may turn to the above-quoted very productive author, W. A. Guy, and we meet here just the same reasoning as on the Continent. In fact, many of these authors choose the same methods independently of each other. In some of his papers he proceeds as Casper or Châteauneuf, calculating the average age at death in some profession or class of society, as, for instance, the Peerage.[1] Sometimes other methods are recommended. He explains (l.c., p. 69) that there

"are two principal means of determining the influence of professions and employments on longevity; the one by comparing the age at death of persons following these professions and employments with the age of death of a class placed in the most favourable circumstances; the other by instituting a similar comparison between the ages of the living."

Thus in a paper on the health of nightmen[2] he states that "the greatest age of any man at work, among the 96 scavengers, was 66, the oldest bricklayer's labourer was 64, and the oldest brickmaker 68."[2] The method was used by others, as, for instance, Thackrath, who wrote a frequently-quoted book on health in various occupations.[3] And Parent-Duchâtelet recommends the

---

[1] W. A. Guy, "On the Duration of Life among the Families of the Peerage and Baronetage of the United Kingdom" (*Journ. Stat. Soc. London*, VIII, 1845).

[2] W. A. Guy, "On the Health of Nightmen, Scavengers and Dustmen" (*Journ. Stat. Soc. London*, XI, 1848).

[3] C. Turner Thackrath, *The Effects of the Principal Arts, Trades and Professions . . . on Health and Longevity*, London, 1831.

method in a paper on the health of workers in the tobacco-industry.[1]

Guy does not, however, confine himself to the two chief methods which he explained. Thus he collects observations on out-patients in King's College Hospital[2] with regard to the relative frequency of phthisis in various professions. Or he deals with the causes of death in London, particularly as regards this disease. In some of his investigations he consulted the well-known actuary F. G. P. Neison, whose contributions will be quoted below; apparently Neison did not object to Guy's material. Though Neison himself made several investigations on a thoroughly correct basis he was liable to make serious mistakes, as it appears from a frequently quoted essay on the influence of intemperance on health.[3] In fact, he had only observations on deaths at his disposal here without knowing the number of persons exposed to risk. The statisticians of those days did not always clearly discriminate between defective observations and reliable material.[4]

It was of course quite natural that the medical statisticians of those days should so often turn to observations in hospitals, lunatic asylums, etc., when trying, for instance, to find the average age of death in various occupations, or the relative frequency of some disease, such as phthisis, compared with other causes of death.

This sometimes led to more or less complicated calculations. This applies to Fuchs, who, in 1835, published a work on the influence of employment on

---

[1] Parent-Duchâtelet et d'Arcet, "Mémoire sur les véritables influences que le tabac peut avoir sur la santé des ouvriers occupés aux différentes préparations qu'on lui fait subir" (*Annales d'Hygiène publique*, 1, 1829).

[2] W. A. Guy, "Contributions to a Knowledge of the Influence of Employments upon Health" (*Journ. Stat. Soc. London*, VI, 1843); "Further Contributions . . ." (VI, 1843); "A Third Contribution . . ." (VII, 1844).

[3] F. G. P. Neison, "On the Rate of Mortality among Persons of Intemperate Habits" (*Journ. Stat. Soc. London*, XIV, 1851).

[4] Yet on a later occasion Neison clearly criticises both Casper's and Guy's methods. Neison, *Contributions to Vital Statistics*, 3rd ed., 1857, p. 131.

health.[1] His material consisted in observations from an institution for diseased artisans. He tries to find the numbers of members indirectly by means of their contributions 1818–34. For the preceding years he assumed that the frequency of disease is the same as later. Having found the number of persons exposed to risk, he compares the frequency of disease in varying employments. It is a defect that he deals with the attacks only, not with the duration of illness, and the numbers are often very small.[2] He is well aware that his observations are not unbiassed, many prosperous artisans preferring to be treated at home. Nevertheless he draws the conclusion that the poor artisans are more frequently taken ill than those who are well paid.

56. As remarked above, it would be easy enough to draw up a long list of erroneous conclusions arrived at in this period. But it is more satisfactory to find proofs of a growing insight into the problems, and it is evident that the period concerned, with all its short-sighted methods, shows constant progress. Even though erroneous methods often caused great confusion, these energetic investigations brought many undeniable results. In fact, the effect of the causes sometimes was so enormous that even an imperfect method would disclose it. This was, for instance, the case with the appalling mortality in prisons in those days. Here Villermé may be quoted.[3] The same author described the great mortality of foundlings.[4] Several investigations

---

[1] Fuchs, "Ueber den Einfluss der verschiedenen Gewerbe auf den Gesundheitszustand und die Mortalität der Künstler und Handwerker in den Blüthenjahren; nach den Tabellen des Instituts für kranke Gesellen zu Würzburg von 1786–1834" (*Heckers neue Wissensch. Annalen der ges. Heilkunde*, 1835).

[2] "Die 22 Meister, welche nach 1803 mit zahlreichen Gesellen den Haarputz der Bewohner Würzburgs in Ordnung hielten, sind auf 8 herabgesunken, die zusammen nur 5 Gesellen nähren" (l.c., p. 413). Very much the same Character has a treatise by Cless, *Medicinische Statistik der innerlichen Abtheilung des Catharinen-Hospitals zu Stuttgart in seinem ersten Decennium 1828–1838*, Stuttgart, 1841.

[3] Villermé, "Note sur la mortalité parmi les forçats du bagne de Rochefort" (*Annales d'Hygiène publique*, 5, 1831).

[4] Villermé, "De la mortalité des enfans trouvés" (*Annales d'Hygiène publique*, 19, 1838).

on the health of *soldiers* and *sailors*, particularly in tropic parts of the world, can also be quoted.[1] The great frequency of *suicides* among troops was one of the facts disclosed by investigations of this kind, as also the high rates of mortality of troops in war times, from disease (particularly epidemic disease). On the whole, epidemics formed a vast field for investigation. One of the authors to be mentioned in this respect was F. Bisset Hawkins, who in 1831 described the wanderings of the *cholera*.[2]

Less successful were the efforts to trace the effects of more complicated causes. Thus the problem of the mortality in various *professions* was in the period concerned and many years after subject to great confusion. The influence of *race* on health was also a difficult problem, as here so many causes co-operate (economic conditions, hygiene, etc.). The Prussian statistician Hoffmann tried in vain to solve the problem with regard to the *Jews*.[3] The same can be said of the attempt of Sykes concerning the health of *Mahomedans, Hindoos and Armenians* in Calcutta.[4]

Progress in method may be observed in various ways. Thus we find a method of *standard calculation* by investigations regarding mortality recommended by F. G. P. Neison, who applies the method to various districts of London with the age distribution in Bethnal Green as

[1] In France, for instance: Boudin, "Études d'hygiène publique sur l'état sanitaire et la mortalité des armées de terre et de mer" (*Annales d'Hygiène publique*, 36, 1846). "Études sur l'état sanitaire et la mortalité de l'armée" (l.c., 42, 1849).

In England: Tulloch and Sykes, thus:

Tulloch, "Comparison of the Sickness, Mortality, and prevailing Diseases among Seamen and Soldiers as shown by the Naval and Military Statistical Reports" (*Journ. Stat. Soc. London*, IV, 1841).

Sykes, "Vital Statistics of the East India Company's Armies in India, European and Native" (l.c., X, 1847).

[2] F. Bisset Hawkins, *History of the Epidemic Spasmodic Cholera of Russia*, London, 1831.

[3] J. G. Hoffmann, "Betrachtungen über den Zustand der Juden im Preussischen Staate" (*Sammlung kleiner Schriften staatswirthschaftlichen Inhalts*, Berlin, 1843).

[4] Sykes, "On the Population and Mortality of Calcutta" (*Journ. Stat. Soc. London*, VIII, 1845).

standard.[1] Methods of this kind appear also in economic statistics, as, for instance, in a lecture by Wm. Neild on the income and expenditure of certain families of the working classes (*Journ. Stat. Soc. London*, IV, 1841), where prices in 1836 and 1841 are compared by using the same consumption as standard. A short method of comparison which may be said to contain the germ of methods used much later, was recommended by Quetelet.[2] If statistical results are found for various districts we may in each case arrange the districts according to the size of the quantity which has been ascertained. Further, the numbers given to a district in the various series may be added and the result will give a comprehensive idea of its more or less favourable condition. During that period, methods of this kind were, however, very rarely used.

The most conspicuous progress was perhaps made with regard to mortality-tables. In the long run the usual method of distributing the deaths per mille according to age could not be satisfying. The problem was to find reliable data for the numbers of persons exposed to death, and here the census was in many cases neither detailed nor accurate enough. Euler's method of reducing the numbers of deaths in the various ages in order to take the increase of population into consideration, was mentioned above (art. 25). A similar method was applied by David Jones.[3] Tellkampf's life-table for Hanover was based on the same principle[4]; in spite of imperfect interpolation it seems to have given fairly reliable results, at least up to the age of 70. In France, Demonferrand made very interesting calculations, using the statistics of conscription for military

[1] F. G. P. Neison, "On a Method recently proposed for conducting Inquiries into the Comparative Sanatory Condition of various Districts" (*Journ. Stat. Soc. London*, VII, 1844).

[2] Quetelet, "Sur le recensement de la population de Bruxelles en 1842" (Extrait du Tome I du *Bulletin de la Commission Centrale de Statistique de la Belgique*, 1843).

[3] David Jones, *On the Value of Annuities*, 1843.

[4] Tellkampf, *Die Verhältnisse der Bevölkerung und der Lebensdauer im Königreich Hannover*, 1846.

service as a means of correction.[1]   His life-table seems to be reliable, at least for males under 20–21, whereas it is difficult to judge with certainty for other ages.

But in this epoch the *census* began to present itself as the easiest way to finding the number of persons exposed to death.   William Farr constructed a life-table (1843) on basis of the census 1841 and the deaths in the same year.

Later followed Life-Table No. 2, on the census 1841 and the deaths 1838–44.[2]   The Danish mortality statistics were treated by E. Fenger (1842), who calculated a mortality-table on the basis of the census 1834 and 1840, and the deaths 1835–9, with a later extension to the period 1840–4.[3]   For the first years of age he used the births instead of the census-results, though without taking notice of the distribution of deaths of infants, according to birth-year.   It is interesting to see how he in the course of his calculations becomes aware of the difference between intensity of mortality and probability of dying within a year.[4]   Mortality in *Iceland* was dealt with in an able book by Schleisner.[5]

In *Belgium* most mortality-tables were calculated on deaths only.   Ducpetiaux, however, constructed a table for Brussels on the basis of the deaths in 1840–2 and the census of 1842.[6]   Quetelet confined himself at first to calculations on deaths only, but finding the census of 1846 sufficiently reliable, he calculated a table on the basis of this census and the deaths of 1840–50.[7]

---

[1] Demonferrand, "Essai sur les lois de la population et de la mortalité en France," *Journal de l'école polytechnique*, 1838 et 1839.   About the results he remarks: "On y parvient après quelques tâtonnements qui n'offrent d'autre difficulté que la longueur des calculs," l.c., 1838, p. 280.

[2] 9th Report of the Registrar-General of Births, Deaths and Marriages in England (1849); see William Farr, *Vital Statistics*, p. 479.

[3] E. Fenger, "Om Dødelighedsforholdene i Danmark" (*Det kongelige medicinske Selskabs Skrifter*, 1848).

[4] l.c., pp. 30–1.

[5] P. A. Schleisner, *Island undersøgt fra et lægevidenskabeligt Synspunkt*, København, 1849.

[6] *Bulletin de la Commission Centrale de Statistique*, II, Bruxelles, 1845.

[7] Quetelet, "Nouvelles tables de mortalité pour la Belgique" (*Bull. de la Commission Centrale de Statistique*, IV, 1851).

The experience of the *English friendly societies* provided an excellent addition to these calculations. Charles Ansell used experience of this kind for the five years January, 1823–January, 1828. Unfortunately the material was small—only about 24,000 years of life, nevertheless this report is a good contribution to our knowledge of the health of provident classes in those days.[1] Several years later F. G. P. Neison embarked upon similar investigations and succeeded in getting a very large amount of material from friendly societies through a system of prizes awarded for the best records.[2]

Some years later a report was made by Henry Ratcliffe on the experience of the Manchester Unity of Odd Fellows[3] and a few years later a weighty report by A. G. Finlaison.[4]

These were not the only investigations on health questions. An interesting although not particularly successful attempt at *retrospective* mortality statistics was made by Daniel Griffin, in Limerick.[5] Every parent who applied for medical relief at the dispensary was questioned for particulars regarding deaths and causes of deaths of parents and children.

A much more solid basis was used by the *life insurance societies*. The experience of the old life-office, *The Equitable*, was treated—rather imperfectly—by A. Morgan (1834). Later followed a report by Galloway on

---

[1] Charles Ansell, *A Treatise on Friendly Societies* (published under the Supervision of the Society for the Diffusion of Useful Knowledge, London, 1835).

[2] F. G. P. Neison, "Contributions to Vital Statistics, specially designed to elucidate the Rate of Mortality, the Laws of Sickness and the Influences of Trade and Locality on Health, derived from an extensive Collection of Original Data, supplied by Friendly Societies and proving their too frequent Instability" (*Journ. Stat. Soc. London*, VIII, 1845; IX, 1846). To a great extent Neison's investigations were incorporated in his *Contributions to Vital Statistics*, 1st ed., 1845; 2nd ed., 1846; 3rd ed., 1857. On the history of Morbidity Statistics cf. Rusher, "The Statistics of Industrial Morbidity in Great Britain" (*Journ. Stat. Soc. London*, LXXXV, 1922).

[3] Henry Ratcliffe, *Observations on the Rate of Mortality and Sickness existing amongst Friendly Societies*, Manchester, 1850.

[4] A. G. Finlaison, *Friendly Societies, Sickness and Mortality*. Ordered by the House of Commons to be printed, 16 August, 1853, and 12 August, 1854.

[5] Daniel Griffin, "Enquiry into the Mortality occurring among the Poor of the City of Limerick" (*Journ. Stat. Soc. London*, III, 1840).

The Amicable.[1] Still more valuable was the co-operation of seventeen English life-offices (including the two mentioned here). The co-operation was arranged in 1838; the report, by Jenkin Jones, on the results was published in 1843.[2] Though the material was not quite homogeneous, some observations being on policies, others on individual lives, and though any life insured in two offices was counted twice over, this report undoubtedly was a long step forward. Various problems of selection were considered; thus E. J. Farren made investigations on mortality according to the ages of the policies.[3]

57. The progress was not confined to mortality statistics, though it was less conspicuous in most of the other branches of statistics. In *birth statistics* it was an interesting problem to discuss the distribution per month of conceptions or births. Of course the interpretation of the results might be dubious. A couple of generations before Wargentin had drawn the conclusion that opulence and luxury were a hindrance to fertility. Villermé took another standpoint, that bad and insufficient food will cause a decrease in the number of conceptions.[4]

As we have seen, the *balance of the two sexes* was studied very early. In the nineteenth century earnest efforts were made to discover some of the causes of this balance. Here, however, a great difficulty appeared, namely, that the solution of the problem required a large number of observations, whereas usually there was very little material to draw upon. This obstacle, conspicuous in this period, remained so for many years after. The most popular theory of the time pointed to the

[1] Th. Galloway, *Tables of Mortality deduced from the Experience of the Amicable Society for a perpetual Assurance Office during a Period of 33 Years ending April 5, 1841*, London, 1841.

[2] *A Series of Tables of Annuities and Assurances calculated from a New Rate of Mortality amongst Assured Lives, with Examples*, London, 1843.

[3] E. J. Farren, *Life-Contingency Tables*, Part I. The Chances of Premature Death, and the Value of Selection among Assured Lives. London, 1850.

[4] Villermé, "De la distribution par mois des conceptions et des naissances de l'homme" (*Annales d'Hygiène publique*, 5, 1831).

difference between the ages of the parents as a significant cause of deviation from the usual average.   If the father was younger than the mother, according to the theory, the female births would be relatively more frequent than in the opposite case.   The theory was maintained in Germany by J. D. Hofacker,[1] and in England by M. Th. Sadler,[2] who worked independently of each other.   Hofacker based his theory on only about 2,000 observations.   Sadler's material—from the British Peerage—had about the same dimension; moreover, he made some puzzling observations concerning widowers who were said, in their second marriage, to beget relatively many daughters.   Sadler's theory, however, was only an appendix to his discussion on the population questions; his theory here being that fecundity is decreasing when the density of population is increasing.[3]   On the whole, Sadler was more of a social reformer than a statistician.

Marriage statistics also made progress.   A curious effort to find the age distribution of the married was made by Quetelet on the basis of the census.[4]   If, for instance, we know the marital condition of the males enumerated at the census, at the ages of 16–20 and 20–25, say 96 and 3,278 respectively, we may try to find the yearly number of marriages at 20–25.   Supposing that the population is constant, with no variations in the age groups, the number of marriages at 20–25 years of age will be 3,278, less the number of those

---

[1] J. D. Hofacker, *Dissertatio de qualitatibus parentum in sobolem transe untibus* . . . 1826, later in German under the title, *Ueber die Eigenschaften, welche sich bei Menschen und Tieren von den Eltern auf die Nachkommen vererben*, Tübingen, 1828.

[2] M. Th. Sadler, *The Law of Population*. A Treatise in six Books, in Disproof of the Superfecundity of Human Beings, and Developing the Real Principle of their Increase. London, I–II, 1830.

[3] The prolificness of marriages is everywhere regulated by the state of the population, and is, *cæteris paribus*, the greatest where the inhabitants are the fewest on a given space (l.c., II, p. 359, see above, art. 50).   Further reference as to the literature on the sex proportion in Wedervang, *Om Seksualproporsjonen ved Fødselen*, Oslo, 1924.

[4] A. Quetelet et Ed. Smits, *Recherches sur la reproduction et la mortalité de l'homme aux différens âges et sur la population de la Belgique*, Bruxelles, 1832.

who already entered marriage before 20. Of these 96 were counted, but a certain number having died afterwards, the reduction is supposed to be only from 3,278 to 3,187, the average number of marriages thus being 637. He then proceeds in the same way from quinquennium to quinquennium. These calculations may be open to criticism as to the suppositions with regard to mortality, but it is interesting to meet with an attempt of this kind to find the movements of population by the help of the census.[1] Of course, such methods will be obsolete when vital statistics give direct observations on the age distribution.

As to the *physical* and *moral* qualities of men, several contributions appeared in this period. Most successful were the efforts to measure the *physical* qualities, as, for instance, the height of persons conscripted for military service. Here the distribution of the observations around the mean was a good illustration of the law of error. In this chapter of anthropometry Quetelet's contributions deserve mention.[2] But other authors also made investigations of this kind, as, for instance, the Danish mathematician, Andræ.[3]

Quetelet did not confine himself to anthropometric questions of this kind. Early in his statistical writings he deals with the correlation between age and height, and tries to find a mathematical formula for it.[4] It may, however, be objected, that there may be a selection which may influence the height, if, for instance, small individuals have a high rate of mortality. On the whole Quetelet was too much inclined to attribute

---

[1] Westergaard, *On the Study of Displacements within a Population.* Quart. Public. of the American Statistical Association, XIX, 1920.

[2] Quetelet, *Recherches statistiques Bruxelles*, 1844. Lettres sur la théorie des probabilités appliquée aux sciences morales et politiques. Bruxelles, 1846.

[3] Andræ, "Bemærkninger ved Høidelisterne over det værnepligtige Mandskab i 1ste sjællandske District" (Appendix to Thune, *Om den fuldvoxne værnepligtige Bondeungdoms Legemshøide i Danmark. Det kongelige medicinske Selskabs Skrifter.* Ny Række, København, 1848).

[4] Quetelet, *Sur l'homme et le developpement de ses facultés, ou essai de physique sociale*, I–II, Bruxelles, 1836 (the quotations are after this edition; the original edition appeared in Paris, 1835).

the character of a physiological law to his results, even though all disturbing causes had not been eliminated.

58. Much more dubious than these anthropometric results were the conclusions in the so-called *"moral statistics."* This expression was introduced by the French lawyer, A. M. Guerry, to whom the Academy awarded a prize for a work on this subject.[1] He dealt particularly with crimes, but his definition also embraces other subjects, such as suicides, illegitimate births, charity and education.

Like the previously mentioned investigations in England by Neison and others, works of this kind could only indirectly throw light upon human morals. Guerry compares the numbers of accused persons in the various parts of France with the population. He also tries to find the correlation between the numbers of persons conscripted for military service who are able to read and write, and the numbers of accused persons who have learnt to read. He noted the great frequency of recidivism, particularly in the case of crimes against property.

Guerry's reputation was so far exceeded by the fame of Quetelet, who in his chief work, *Sur l'homme*, tried to apply statistical methods to all branches of human life. In the period concerned, he is the central figure in statistical literature, and his contributions show both the strength of this period as well as its weaknesses. Anxious as he was to extend statistical investigations to the widest possible limits, he was naturally tempted to accept even badly-founded results without much criticism (for instance, on mortality).

As an illustration of this sort of light-hearted investigation, his attempt at finding the correlation between poetical production and age can be mentioned.[2] He collected observations on 47 French authors who have

---

[1] A. M. Guerry, *Essai sur la statistique morale de la France.* Ouvrage qui a obtenu le prix de statistique décerné en 1833 par l'académie royale des sciences. Paris, 1833. Many years later he again won a prize for his *Statistique morale de l'Angleterre comparée avec la statistique morale de la France*, 1864.

[2] *Sur l'homme*, II, pp. 119 *sq.*

written 165 dramas which are classified somewhat arbitrarily according to their value. The dramatic talent,
according to his statistics, manifests itself after the
twenty-first year of age; it seems to be very vigorous
between 30 and 35, and it is growing up to 50–55;
after that age there is a conspicuous decrease, particularly in the value of the dramas.

Better known are his descriptions of the *criminal man*.
Already in a paper of 1831 [1] he speaks of a tendency
towards crime (*penchant au crime*). Evidently leaving
out of consideration that there are different types of
criminals each with its own propensity to crime, he
draws a picture of a person who is tempted by all sorts
of crime: the bent to theft begins early in life, mani-
festing itself at first in the home, later out of the home,
and in the end on public highways, and maybe combined
with violence; still earlier is the bent towards sexual
crimes, changing as it does by and by, to end with
the murder of the victims. The last steps of the
career of the criminal are characterised by falsehood,
as a compensation for the brute-force of his younger
years.

This theory again leads to Quetelet's famous theory
of *l'homme moyen*, to whom he attributes all the qualities which the statistical observations disclose as typical.[2] This typical man has the average stature, weight,
etc., and the same holds good with regard to all intellectual and moral qualities. He is in human society
the same as the centre of gravity of a physical body,
the point around which all social elements move. Moreover, Quetelet considers this average man as a *type of
beauty*, and this he even extends to moral qualities, the
beauty consisting of a just balance between the facul-

---

[1] Quetelet, *Recherches sur le penchant au crime aux différens âges*, Bruxelles,
1831. His theory is repeated in later publications; for instance, *Sur l'homme*,
II, pp. 248 *sq.*

[2] Quetelet, *Recherches sur la loi de la croissance de l'homme*, 1831 (Nouveaux
mémoires de l'académie royale des sciences et belles lettres de Bruxelles, VII,
1832). See also *Sur l'homme*, as well as his later work, *Du système social et
des lois qui le régissent*, Paris, 1848.

ties. An individual who had all the qualities of the average man would represent all that exists of grand, beautiful and good.[1]

This theory of an average man may be useful as long as we are confined to physical qualities. A sculptor may ask what are the average measures of the various limbs of a human body, and in most cases he will consider a conspicuous deviation from the mean, ugly, as, for instance, a short arm. But it seems that Quetelet failed to see that it would be impossible to construct a statue with absolute exactitude from all the various averages, these averages being only a guide to the artist and nothing more.[2] And it is evident that it is still more difficult to apply average intellectual or moral qualities as characterising an average man, even though these results may be completely well founded. Nor was Quetelet always quite clear in applying his theory; for instance, when distinguishing between apparent and real propensity to marriage, the former being the only one which can be observed statistically. Just as a gambler may have a good chance to win and yet may lose incessantly, a man may have a real propensity to marriage without ever marrying.[3] This is evidently a limping comparison.

As we have seen, *regularity* in the social phenomena had already struck the political arithmeticians in the eighteenth century. Quite naturally the same appears in the epoch here concerned. As Süssmilch believed in the permanence of types, so did also Quetelet. According to him the type of beauty did not change much, the deviations from the mean only might be reduced. As late as in 1871 he submitted the theory that mortality is essentially constant. Curiously enough he finds a proof in the age at death of sixty famous men from various periods of history. That five out of these died

---

[1] *Système social,* p. 277. *Sur l'homme,* II, p. 289.

[2] Thus the average of the sides of rectangular triangles will generally not form a new rectangular triangle

[3] *Système social,* pp. 77 *sq.*

at 35–40 years old was a sign to him that this age was particularly dangerous.[1]

The stability of statistical results caused a vivid discussion on the important problem of the *free will* of man. Whereas authors of Süssmilch's type mostly saw statistical problems in the light of Divine providence, the nineteenth century was more inclined to take the standpoint of naturalistic philosophy. Quetelet speaks of a *physique sociale*; he considered the statistical phenomena as products of physical laws, as results of environment, the human will having no part in the events. Society contains all germs of the coming crimes, the guilty person being only an instrument of execution.[2]

According to Quetelet and his followers, things are just as if a people in a kingdom had resolved to have in each year the same number of marriages, distributed in a definite manner among provinces, among cities and rural districts, among bachelors and widowers, etc.[3] The regularity is even greater than it would be expected according to the calculus of probabilities. A population which consisted only of wise men will show very small deviations from the average, whereas persons who are ruled by their passions are the prey of outer causes, which explains the apparent paradox that social phenomena, under the influence of the *liberum arbitrium*, from one year to another show a greater regularity than where the causes are material and accidental. He maintains that human free will is without sensible effect if the observations are extended to a large number of individuals.[4]

Quetelet did not prove, however, that the phenomena in moral statistics were particularly regular; but even

---

[1] *Véritablement fatal.* See *Anthropométrie ou mésure des différentes qualités de l'homme*, Bruxelles, 1871, pp. 380 *sq.*

[2] *Sur l'homme*, I, p. 10.

[3] *Système social*, p. 67. The same idea will be found many years later in Ad. Wagner, *Die Gesetzmässigkeit in den scheinbar willkürlichen menschlichen Handlungen vom Standpunkte der Statistik*, Hamburg, 1864.

[4] l.c., pp. 96 *sq.*, p. 70.

if this should be the case, deviations from the mean would take place leaving the free will a space for action. In reality the problem of the *liberum arbitrium* could not be solved by statisticians.[1]

Although Quetelet overshadowed most statistical contemporary authors by his brilliant style, his vivid imagination and his abundance of ideas, several other writers may be consulted by the student of his time. Moser's book on the laws of the duration of life[2] gives a good view of the standpoint of statistics a century ago, and the same holds good of a work by Chr. Bernoulli, who with a real critical sense reviews the statistical literature of this period.[3]

The statisticians of the following age started out with a good inheritance. On one side, a well-founded knowledge in important chapters of human statistics. On the other, it was necessary to cut off loose arguments and hasty results which in a period of enthusiasm had a natural excuse but which could not be maintained if statistics were to undergo a calm intensive revision and thereby secure its rank as a science.

It may be that statistics had reached too high a position on the Parnassus of science. The statisticians of those days are not to be blamed for it. But even if they had to go a step downwards it cannot be denied that statistics still had a very rich inheritance; not only in the form of positive results but also in the large number of problems appealing for a solution.

In practical respects official statistics needed better organisation; an extension was required, not only in economic but also in vital statistics; the questions in the schedules had to be put and answered as distinctly and accurately as possible. Further methods needed to be improved, for instance, in mortality statistics.

---

[1] The literature on Quetelet and his time is very numerous; here only two works will be quoted: Frank Hankins, *Ad Quetelet as Statistician*, New York, 1908; Lottin, *Ad Quetelet, Statisticien et Sociologue*, Paris, 1912.

[2] L. Moser, *Die Gesetze der Lebensdauer*, Berlin, 1839.

[3] Chr. Bernoulli, *Handbuch der Populationistik oder der Völker- und Menschenkunde nach statistischen Ergebnissen*, Ulm, 1840–1.

The bridge from the calculus of probability to statistics had to be built stronger and broader. We shall see in the following chapters how far this goal was reached in the decades after the era of enthusiasm.

CHAPTER XIV

STATISTICAL CONGRESSES

59. In the history of the nineteenth century, congresses
might form the subject of an important and attractive
chapter.   In spite of the numerous wars after the middle
of the century, there was evidently a general yearning
for peaceful co-operation between the nations.   An un-
mistakable sign of this was the free-trade movement
which had many supporters and which at this time could
show conspicuous triumphs, especially in the commercial
policy of England.   World-exhibitions indicated a similar tendency.

Small wonder that international co-operation was particularly desired by statisticians.   A mass of statistical
material was now regularly collected, even though official
statistical services were in many countries still very de-
fectively organised.   But there was a lack of uniformity,
which made comparison between the results from various
countries difficult or even impossible, and the exchange
of publications from one country to another was often
irregular and expensive.

The first international statistical congress was held in
Brussels, 1853.   In the following twenty-three years eight
other congresses took place.   In 1855 the second congress was held in *Paris*; two years later came the Vienna
congress 1857, and again after three years the congress in
London 1860.   The next three congresses were in *Berlin*
1863, *Florence* 1867, and *The Hague* 1869.   Finally came
the eighth congress in *St. Petersburg* 1872, and the ninth
in *Budapest* 1876.   In *St. Petersburg* a *Permanent Commission* was created which to some degree was to replace
the congress.   It met four times: in *Vienna* 1873, in

Stockholm 1874, in *Budapest* 1876, and finally in *Paris* 1878, after which conference its work came somewhat dramatically to an end.

It will be worth while to follow the transactions of these congresses in some detail. In doing so we get at the same time, a picture of the progress of official statistics during a couple of decades. It may, of course, be doubtful how far these congresses influenced the progress of official statistics, as there were other paths open to its evolution. The history of statistics shows how already in the eighteenth century political arithmeticians got into touch with each other through correspondence, and even though scientific journals were not as numerous then as now, they were a useful means of communication. The international exchange of statistical publications was slow compared to the present times, but the statistical bureau of a country was not prevented from getting news from abroad as to what was going on there. Wherever intellectual work is done conscientiously in a country, progress will be the result, even though the work be isolated. At all events it seems probable that the statistical congresses assisted this progressive movement. It is interesting to see how anxious the leaders of the first congresses were to get reports from the delegates, of the condition of statistical service in their respective countries. These congresses were held under government auspices; the leaders of the statistical service were sent as official delegates, and even though the governments were not bound to accept the resolutions, indefinite and vague as they sometimes were, they were nevertheless supposed to take an interest in the transactions of the congresses. Besides the delegates, there were several others who took part in the congresses. Many of these were outsiders who only occasionally were connected with statistics. This —as we shall see—was a weak point, contributing to the degeneration of the congresses, but the effect of spreading the interest in statistical problems in the country where the congress was held, may have been one of the circumstances which promoted the evolution of official statistics.

60. The initiative leading to the international statistical congress seems principally due to Quetelet, then on the height of his fame. He had for many years been a welcome guest to statistical circles in England. At the second meeting of the British Association for the Advancement of Science, 1832, it was resolved to form a statistical section. The following year at the meeting in Cambridge a permanent committee of the section was formed to regulate its affairs. Among the members were Malthus, Babbage, John Lubbock and Quetelet, the latter having been sent to the meeting as a delegate from the young *Belgian* Government. Quetelet found the programme too limited and proposed the formation of a statistical society in London, which—as we have seen—was realised in 1834.[1] He could thus justly feel at home in England, when, seventeen years later, he took the initiative towards an international statistical congress. The preparatory discussions with other statisticians took place in London, 1851, during the famous World Exhibition there. According to the original plan, the congress was to be held in 1852, but for political reasons it was postponed till the following year. Quite naturally Brussels was chosen as meeting-place; the Belgian Central Commission undertook the great preparatory work. The Congress was held under the presidency of Quetelet. About 150 members met, many of them being delegates from their respective governments or from public or private institutions.[2]

In many respects this first international statistical congress was the model of the following eight congresses with all their conspicuous merits and defects, and it may therefore be useful to review the report in some detail.[3]

The chief object of the congress was a practical one,

---

[1] Fr. J. Mouat, *History of the Statistical Society of London*, Jubilee Volume of the Statistical Society, June, 1885, p. 15.

[2] See a letter from Ch. Babbage in Report of the Proceedings of the Fourth Session of the International Statistical Congress, held in London, July 11, 1860, and the five following days (London, 1861), pp. 505 *sq.*

[3] *Compte rendu des travaux du congrès général de statistique réuni à Bruxelles les 19, 20, 21 et 22 Septembre, 1853.* Bruxelles, 1853.

viz. to promote the organisation of official statistics and to unify the reports from the various statistical institutions so as to make the documents comparable. The desirability of a common weight and measure was expressed, here, as at subsequent congresses, and a cheap and easy exchange of statistical publications was further recommended. In the resolution accepted with regard to the organisation of a statistical service we may find traces of *Belgian* influence in so far as the system recommended was very much the same as that in *Belgium*:

"That in every State a Statistical Central Commission, or similar institution, be established, consisting of representatives of the principal Government departments and other influential persons, who by reason of their special qualifications are in a position to promote the growth of the study of statistics, and to lend their assistance in the solution of complex problems" (form accepted in Paris, 1855).

This may be read as an attempt at decentralising statistics, but in point of fact the Central Commission was established to facilitate centralisation of all the information hitherto collected by the different administrations, and which had been subject to much casual irregularity. In reality the resolution did not prescribe any definite form of organisation, so long as the practical aim, to collect reliable statistical information, was reached. The tendency of organisation in the period concerned was chiefly to establish central statistical bureaux, whether central commissions were added or not. In reality the resolution did not enter into the principles of organisation.

The programme was not, however, confined to the problem of organisation. It covered, in fact, the whole field of official statistics, and the result of the transactions was a long list of more or less detailed resolutions. It would seem surprising that the congress managed to get through all this in the four days allotted to the transactions, had it not been for the careful preparatory work for which it was indebted to the Central Commission. The programme which was to be laid before the members was first discussed in various sections and was thereafter quickly passed without much further discussion.   The proceedings of the congress were characterised by punctuality and brevity.   The first meeting of the congress with the opening address, etc., and with reports from various delegates, was over in three hours.   But it is evident that the congress was at times under hard pressure, especially on the last day, in order to finish the whole work in due time.

The aim of the congress being chiefly a *practical* one, there was no room for lectures on special *scientific* problems. *Social* problems were on the whole excluded, but it was not possible entirely to keep within that limit.   Thus in one of the sections the problem of family budgets of the working classes was treated.   It will be remembered that members of *English* statistical societies made great efforts to obtain statistical observations as to the condition of the poor, and naturally these investigations would stimulate similar studies in other countries.   In discussing this question the members of the congress could hardly help entering into social questions, though they were aware that they were leaving the neutral ground assigned to the congress.   A resolution was passed recommending that another congress should be held, consisting of persons from various countries, engaged in researches concerning the physical, moral and intellectual progress of paupers and the working classes.

The resolutions of the congress chiefly insisted on statistical information from the whole field of observation, as, for instance, a general census embracing the total population, complete lists of marriages, births and deaths.   Sometimes the resolution recommended very detailed information, as, for instance, a "cataster," a sort of Domesday-book, based on a careful trigonometrical survey and with estimates on the value of property, leaving space for the registration of future changes. The statistics of family budgets form an exception to the rule of covering the whole field, for here a representative method ("sampling") would be natural.

One of the resolutions passed recommends a *general register* of the population in each commune, each family being allotted one page where future changes might be recorded. A general *census* should be taken every tenth year in December; name, age, sex, language, religion, profession, etc., were to be registered on the schedules. The congress distinguishes between the actual population ("population de fait") and the legal one ("population de droit"), preferring the former—in good harmony with the general evolution of official statistics. A rather complicated problem, that of a uniform nomenclature of the causes of deaths, was also under discussion, but the congress referred it to the future.

61. In many respects the following congress, in *Paris* 1855, resembled the first one. The official character was even more emphasised than in *Brussels*, as will be apparent from the lengthy title of reference to the report.[1] As in Brussels, there was no room for special lectures on scientific problems, the congress following its practical aim. As in Brussels, *French* was the only official language, later the language of the country in which the congress was held was allowed together with *French* (*German* in *Vienna* 1857, *English* in *London* 1860). At last in *Budapest* 1876, *German*, *English*, *French*, *Italian* and *Hungarian* were allowed. The question of the official language was, of course, not indifferent, for in *Paris* and *Brussels* members with French as their native language had an advantage, whereas on the other hand the use of several languages would often have caused confusion or loss of time.

As in 1853, the *Paris* congress took much interest in the collection of statistical material; with regard to the questions of cretinism and imbecility the report (Boudin) enters upon several circumstances which may be of influence, such as heredity or geological conditions. As

---

[1] *Compte rendu de la deuxième session du congrès international de statistique réuni à Paris les* 10, 12, 13, 14 *et* 15 *Septembre,* 1855. Publié par les ordres de S. E. M. Rouher, Ministre de l'agriculture, du commerce et des travaux publics, président du congrès, par les soins de M. A. Legoyt, chef du bureau de la statistique générale de France, secrétaire du congrès, Paris, 1856.

in 1853, the social question is touched upon, and it was difficult to remain on neutral soil; thus with regard to night labour in workshops, opinions here were strongly divided. One of the members protested energetically against a discussion of this kind, finding it idle; he would confine the work of the congress to mere facts.[1]   But this social interest could not be removed.

In *London*, 1860, a resolution is passed that the "Congress respectfully urges statesmen, political economists, philanthropists and administrators, to study the condition of the working classes, their wants, their resources, and the measures to promote their well-being,"[2] and the congress in *Florence*, 1867, uses still stronger expressions with the idea even of obtaining co-operation between the various charitable institutions.

It was felt as a severe loss at the Paris congress in 1855 that Quetelet was prevented from attending the meetings through illness.   In the summer of 1855 he was attacked by apoplexy, and although he soon recovered and continued publishing several statistical treatises, he never regained his full powers, his literary production being to a great extent a repetition of older works.   He was, however, able to take part in all the following sessions till his death in 1874, though his contributions to the discussion were not important.[3]   In *London* his former pupil, the Prince Consort, opened the congress with a speech, in which he proved that he had not forgotten what he had learnt from Quetelet in Brussels in 1837 about the calculus of probabilities.[4]

---

[1] "Je désire vivement . . . que le Congrès ne s'égare pas dans des questions oiseuses, tout a fait étrangères à ses travaux, et qu'il soit bien établi, dès à présent, que la statistique ne doit s'occuper que des faits tels qu'ils existent . . ." l.c., p. 333.

[2] Report of the Proceedings of the Fourth Session of the International Statistical Congress, held in London, July 11th, 1860, and the five following days (London, 1861), p. 203.

[3] Lottin, *Quetelet, Statisticien et Sociologue*, Paris, 1912, pp. 85 *sq.*   Frank H. Hankins, *Ad. Quetelet as Statistician*, New York, 1908, p. 31.

[4] A good survey of the third Congress was given by the Austrian statistician Ad. Ficker: *Die dritte Vorsammlung des internationalen Congresses für Statistik zu Wien im September*, 1857.   Wien, 1857.

Comparing the reports of the third and fourth congress with those of the first two, we are struck by the evident progress of official statistics in this short period of seven years. Moreover, the London report with its contributions by Gompertz on the law of mortality, or by Newmarch, on prices and wages, has a more scientific character than in the first congresses. This will, for instance, appear by comparing the resolutions on vital statistics. The London report has a very careful and detailed programme for a census; this congress, further, took much interest in the problem of production and consumption, supporting earlier resolutions on agriculture, etc. Of course, the same degree of clearness could not be expected everywhere. Thus the problem of statistics of literature which had been discussed in *Vienna*, was again, rather confusedly, treated in *London*, where a chaos of items was recommended; it would have been difficult to give more than a superficial treatment of a material of this kind.[1]

The following nine years, with their congresses in *Berlin*, *Florence* and *The Hague*, show continuous progress. As to the theory of statistics, a rather vague resolution was passed in *Florence*, 1867, at the initiative of Quetelet, that there should be created a special section at future congresses, to deal with statistical questions in direct connection with the theory of probabilities. The resolution was passed without discussion. At the following congress, at *The Hague*, Baumhauer enters into a corresponding problem, recommending that statistical investigations should not only deal with averages, but with the deviations from the mean. In discussing this question he also touches the problem of free-will. The resolution which was passed on the recommendation of Engel was

---

[1] A good survey of the first four congresses was given by Engel as a preparation of the congress in Berlin: *Compte-rendu général des travaux du congrès international de statistique dans les séances tenues à Bruxelles 1853, Paris 1855, Vienne 1857, et Londres 1860.* Publié par les ordres de S. E. M. le Comte d'Eulenburg, Ministre de l'intérieur, sous la direction de M. le Dr. Engel, Directeur du bureau royal de statistique à Berlin. Berlin, 1863. Similar surveys were prepared for the following congresses.

followed by an additional one, proposed by G. Mayr, which was equally accepted after a curious discussion about priority, between Engel and Mayr, none of them, however, taking any real interest in the deeper problems of mathematical statistics. As to vital statistics, the congress recommends that in the lists of deaths not only the age but also the birth-year of the deceased persons shall be reported.

In *Berlin*, 1863, problems of a scientific nature also came up for discussion; for instance, those regarding *anthropometry*, although the practical aims of the congresses still had a prominent place. Some of the leading men, as the above-named *Prussian* statistician Engel, felt the difficulty arising from the constitution of the congresses. In Brussels the foreign elements were in the majority, and to a great extent the congress was a meeting of experts, but later the national elements, many of whom were outsiders, were much more numerous. Thus in Berlin there were only 127 foreigners out of 477 members, and in Florence 85 foreigners out of 717 members. It is clear that in such enormous assemblies the value of the proceedings will easily suffer. Neumann-Spallart, who has given a sketch of the history of the statistical congress, remarks, for instance, about the congress at *Florence*, that too much time was taken up by long speeches, and many resolutions were accepted without full discussion.[1] To counteract these drawbacks Engel in *Berlin*, 1863, proposed a scheme for a *permanent organisation* of the statistical congress. This was, however, postponed to a future congress, and it was not till 1872 in St. Petersburg that his plan was realised. In *The Hague* in 1869, a step in that direction was taken at his initiative, a plan of a "Statistique Internationale de l'Europe" being prepared; the work was distributed among several statistical bureaux; unfortunately it made rather slow progress on account of difficulties which arose; still, some positive

---

[1] Neumann-Spallart, *Résumé of the Results of the International Statistical Congresses and Sketch of Proposed Plan of an International Statistical Association.* Jubilee Volume of the Statistical Society, London, 1885, p. 294.

results, for instance, with regard to shipping, may be recorded.

In *Florence*, and later, a so-called *Avant-Congrès* was held, consisting of the delegates, and in *St. Petersburg* finally the *Permanent Commission* was established, chiefly composed of members to whom was to be assigned the duty of preparing international statistics. Under these measures the constitution of the congress grew rather complicated. There was in *Budapest* in 1876, as usual, an organising commission, which had to co-operate with the permanent commission; but as the latter did not embrace all the delegates, it was agreed to let the Avant-Congrès remain in order to prepare the order of business, and decide upon the languages allowed.

62. The proceedings of the last two congresses were by no means devoid of interest. *St. Petersburg* came to an agreement concerning the census (as already in *Brussels*), with distinction between the legal population (*population de droit ou legal*) and the actual one, the latter being chiefly taken into consideration. Registers of population were recommended, also civil registration of births, deaths and marriages, further statistics of prostitutes, elaborate industrial statistics, etc. As at all the preceding congresses, the *St. Petersburg* congress was interested in criminal statistics. A uniform nomenclature of crimes being wanted, the congress found it desirable to have an individual card for each accused person. The ninth congress which met four years later in *Budapest* took a particular interest in *mortality* statistics. There were interesting contributions to the problem of how to get reliable mortality-tables; for instance, by the Hungarian statistician J. Kőrösi, and by K. Becker, at that time director of the statistical office of the German Empire.

In the meantime the new Permanent Commission set to work. Whereas the congress had too many members, here the number was comparatively small. At the first conference, in *Vienna*, 1873, there were only 21 members. The distribution according to nations was, however, unsatisfactory, no English representative, for instance, being present.

It was not very easy to find a medium between the opposing opinions as to the aim of the Commission, the most radical view being represented by Engel, who wished to keep the centre of gravity in the Commission. In time the constitution and object of the Permanent Commission were agreed upon, and after the meeting in *Paris*, 1878, the Commission appeared to have a good horoscope. According to the resolutions agreed upon, the Commission was to consist of representatives of official statistics, delegated by the Governments, of delegates from great cities, or from other institutions, and also of persons specially nominated by the Commission. Its aim was to be a link between the congresses; it was to work for the promotion of international statistics, to call the attention of the organising commission to the questions to be debated at the next congress, and to publish periodical publications (*Bulletin de la Commission permanente du Congrès international de statistique*). It was decided to abide by the decision of the majority of the Governments as to the manner in which effect should be given to the decisions and results arrived at by the commission.[1]

The majority of the representatives of statistical bureaux recommended to their respective Governments approval of these resolutions, but the representatives of the Statistical Office of the *German Empire* were in opposition, and as an invitation to the sitting of the Commission at *Rome*, 1879, was made, some of the *German* statisticians declined to accept this invitation, as well as that to a tenth statistical congress in 1880. So the sitting in *Rome* had to be postponed, and the President, the Hungarian statistician Keleti, resigned. In this way the International Statistical Congress came to an end.

Many statisticians deplored this tragic end of international congress work. But from other influential

---

[1] Neumann-Spallart, *Die vierte Session der Permanenz-Commission des internationalen statistischen Congresses in Paris*, 1878. Statistische Monatschrift, IV, Wien, 1878.

sources it was found that the congress institution had outlived itself, and that the progress made under the influence of the congress was bought at too high a price.

The outward character of the congresses was brilliant, under eager competition between the various nations where the congresses were held, with distribution of stars and ribbons, excursions and banquets for the members, but it is not to be wondered at that the Governments grew tired of this lavish hospitality. Intercourse between statistical experts was now so easy, the scientific journals so open for discussion on statistical problems, that nobody had reason to fear any hindrance to official statistics if the international congress disappeared for ever. Some of the leading statisticians of the time, such as Mouat and Newmarch openly criticised the congresses. Newmarch declared in a discussion on the *Budapest* congress that he could not see that any large amount of practical good had resulted from the statistical congresses, which he characterised as international picnics.[1] If association between the statisticians should hereafter be tried, it was clear that other forms would have to be invented than that of an International Congress. This was realised when some years later the International Statistical Institute was established.

63. Whether the Congresses really contributed to the promotion of official statistics in this period or not, we can at all events record several cases of practical advance in various countries, and independently of these changes we notice a real growth in the inner value of the publications.

In *England*[2] the most outstanding event was perhaps the Act by which, in 1874, registration of births, deaths and marriages was made compulsory; this added greatly to the accuracy of the reports on vital statistics which, under the influence of William Farr, on the whole

---

[1] Fr. G. Mouat, Second and Concluding Report of the Ninth International Statistical Congress, held in Buda-Pesth, in September, 1876 (*Journ. Stat. Soc. London*, XL, 1877, pp. 554–5).

[2] Here, as well as later, apart from numerous monographs, we can refer to the above-quoted Memorial Volume: *History of Statistics*, 1918.

reached a high standard, especially as to the mortality in various professions. In 1853 civil registration was extended to *Scotland*, and about ten years later to *Ireland*. Other statistical work, as, for instance, that of the Board of Trade, continued after Porter's death by Valpy and later by Giffen, deserves mention, not to speak of early efforts with regard to labour statistics (friendly societies). The Labour Department of the Board of Trade was, however, not established till later (1886). Less conspicuous was the progress in the *Colonies* and *India*. In *India* most statistics date from 1881 or later; the first general census of India was taken in that year.

In *Canada* there were great difficulties.[1] A Board of Registration and Statistics was created in 1847 for the province of Canada (then embracing the provinces of Quebec and Ontario), and a census was taken in 1851 and 1860, but the Board soon became very much disorganised, and the census reports appear to have been distinctly inaccurate. It was not till 1868, after the establishment of the *Confederation of Canada* in the preceding year, that a reform was made. The census, statistics and the registration of statistics were then placed under the new federal Department of Agriculture, and a census was taken in 1871 under more favourable auspices. Conditions seem to have been better in *Australasia*; thus in *New South Wales* a general registry of births, deaths and marriages was established in 1856, and a regular decennial census was taken from 1861 onwards. But it was not until 1873 that *Victoria* appointed a "Government Statist," at the same time registrar of births, deaths and marriages, and only in 1886 did *New South Wales* get a statistical office.

In the *United States of America* statistics were principally concentrated upon the decennial census. The census of 1840 appears to have been very defective, but the following census of 1850 showed important progress. In 1840

---

[1] Godfrey, "History and Development of Statistics in Canada," *History of Statistics*, 1918, pp. 181 *sq.* See further, Robert R. Kuczynski, *Birth Registration and Birth Statistics in Canada*, Washington, D.C., 1930.

the family was the unit of enumeration; now the individual, as well as the slaves, were considered and a number of questions were introduced on the schedules.

The mortality schedule required data concerning deaths during the year, including cause of death; the agricultural schedule required detailed returns for each farm; for instance, quantity produced and value of animals slaughtered; there was a schedule for industries, asking for returns of capital invested, quantity and value of production, and the average number of male and female employees. This census seems to have been a success, but it was a serious drawback for continuity that no permanent office was established, a new office having to be created at every census. This was not repaired until the present century. In 1856 a statistical office was created, but this institution had nothing to do with vital statistics, its duty being to prepare annual reports of the commercial relations of the *United States*. Some years later a statistical bureau of agriculture was established. The United States did not get a bureau of labour until 1884; institutions of this kind were, however, established earlier in several single States. Thus in *Massachusetts*, a bureau of statistics of labour was created in 1869, with the object of gathering statistical data relating to the welfare of the wage-earning classes of the population. It is curious that the U.S.A. had no delegates at the first three statistical congresses; it was in Vienna that it was first resolved to invite the U.S.A. to the next congress.

Passing from America to the *Continent of Europe*, we meet in *France* with various attempts at promoting official statistics. As to the census, the enumeration in 1876 was based on individual reports (bulletins) supplementing the household reports; this was a very important improvement. The Bureau de la Statistique Générale which Thiers had created in 1833 was subject to modifications in 1852 and 1871. Statistics of agriculture were classified under this bureau till 1881, but this branch of statistics was unfortunately a stepchild. "The data were furnished by arbitrary and superficial estimates due to the transient collaboration of incompetent administrators, etc. . . ." [1] Official statistics in this period in France were very much decentralised, numerous bureaux having each its particular duty. An interesting institution was the Bureau de la Statistique Graphique, which was created in 1878. It may be looked upon as evidence of the interest taken by the Statistical Congresses in graphic statistics.

In *Belgium* commercial statistics were altered in 1854 in so far as a new classification of the foreign commercial products was adopted. A second industrial census was taken in 1866, though not published; the census of agriculture was repeated in 1856, 1866 and 1880 simultaneously with the enumeration of the population. A law of 1856 sanctioned the regular keeping of population registers. It was a drawback with regard to the census of population that it was very much decentralised, the central bureau having only to transcribe and add the tables which were furnished by communal administrations.

In *Holland* official statistics had not exactly favourable conditions. The statistical bureau which had been established in 1848, under the direction of Baumhauer, only lasted for thirty years. It was abolished in 1878, and, as in *Belgium*, official statistics suffered from decentralisation, several branches remaining with other departments of the general administration. A law of 1850 prescribed the establishment of provincial bureaux of statistics, but these institutions never became an active part of official statistics, several of them disappeared, and at last they were abolished in 1905. An interesting co-operation with a statistical society (formally organised in 1856) went on through a number of years, and as the statistical bureau was abolished, the society obtained a yearly allowance in order to publish various statistical reports. In accordance with the recommendation of the statistical congress at *Paris* in 1855, a Government Commission of Statistics was established in 1858, with the object of bringing about greater scientific unity in statistics.

[1] Faure, l.c., p. 299.

In *Germany* political events in 1866 and 1870 changed the conditions for the organisation of official statistics. An Imperial Statistical Office was established in 1872. Apart from this there were many statistical offices in the various States which formed the German Empire. Several of them were established at this period. Thus the Kingdom of *Saxony*, as mentioned before, got its statistical office in 1851, *Mecklenburg-Schwerin* in 1851, *Oldenburg*, 1855, etc. The free City of *Hamburg* joined in 1866, *Bremen* in 1867, and *Lübeck* in 1871. In some of the States special commissions had the task of bringing about uniformity in the various branches of statistical work, particularly the Central Statistical Commission in Prussia.

This period also saw the inauguration of a number of municipal statistical offices. The oldest one, in *Berlin*, was established in 1862; it soon rose to a prominent position, particularly under the direction of Böckh. Other countries also set in motion a municipal statistical service; for instance, *Budapest* in 1869 (Kőrösi to 1906), *Vienna* in 1862, etc.

In *Austria*, the statistical service was long in arrears. However, in 1840 the Empire established a statistical office. A census was taken in 1857 (embracing also live-stock); a Central Statistical Commission was created 1863. In 1869 a law was passed which modernised the census, in 1877 a permanent commission was appointed to determine yearly the average commercial value of the imports and exports of the past year, and in 1873 a department of statistics of agriculture, forestry and mines was created.

Hungary, too, was behind for a while. Not until after the agreement with *Austria*, in 1867, was an official statistical organisation created, as a department of the Ministry for Agriculture, Industry and Trade; it was made more independent in 1871. A general census was carried out in 1869. As in *Hungary*, so in *Italy*, statistical service was in an unfavourable condition owing to political causes. The Kingdom established a statistical office in

1861, and in 1868 a Central Commission. Under the energetic direction of Bodio (from 1872) official statistics reached a prominent position.

In Russia, natural causes helped to retard any statistical service for a long time. Before 1897 Russia had no real census. The so-called revisions, a sort of registration of the population, served as a basis; they were chiefly of a fiscal-political character, including only persons liable to a head-tax and to military service. They were far from being exact, and moreover they ceased altogether in 1858. In addition to these "revisions" so-called family registers existed, particularly for the purpose of military service, and finally administrative and police estimates of the population. The statistical service was reorganised in 1863 with a Central Statistical Office under the direction of Szemenow, who had a good name as geographer and statistician; he was at the same time chairman of the statistical council. The provincial statistical offices created in 1834, which were in a miserable condition, were reorganised in 1860; but even after this reform they hardly deserved their name.

The statistical office undertook one important investigation, a study of *landed property*, in 1877, in connection with the agrarian reform of 1861. Szemenow used a system of checking the material; for each province the material relating to a single district had to be transmitted in order to be tested; whereupon the provincial officers received instructions as to the continuation and control of the investigation. This single fact will show the enormous difficulties which he had to encounter. The central statistical office also tried to get statistical information as to sowing and harvests in 1860–4, but the results were very unsatisfactory and they were therefore not published.

With all these handicaps it is not surprising that the attempts at *local* statistics should also prove unsuccessful.

Finland [1] was during the greater part of the nineteenth century ruled by *Russia*, but its whole system of statistical

---

[1] Martti Kovero, *Official Finnish Statistics*, Helsinki, 1924.

service received its stamp from *Sweden*, vital statistics, as in that country, resting chiefly on reports from the clergy. Whereas in *Sweden* the lists are nominative, in *Finland* nominal rolls were not prepared. The pastor sent in complete tables for his parish, the Central Statistical Bureau merely compiling the results. Not until 1865 was a statistical Department established; as in Sweden, a statistical Central Commission retained contact between the various branches of official statistics, till in 1884 this duty was committed to the Central Statistical Bureau, as it was henceforward called.

In *Sweden* the most important event was the reorganisation of a statistical service in 1858. A committee appointed by the Government had recommended that the existing tabulating commission should be transformed into a statistical department which should take care of vital statistics and various other branches; there was, however, still a considerable decentralisation, many branches of statistics being entrusted to other departments, and the desired uniformity was to be obtained through an advisory commission similar to that in *Belgium*. In the main this plan was adopted, becoming the foundation-stone to the present organisation.

In *Norway* also a reorganisation of official statistics took place. The statistical bureau of 1837 was in 1875 transformed into a more independent institution under the name of the Central Statistical Bureau. Vital Statistics made considerable progress in this period. Thus, in 1865 a very important step forward was taken with regard to the census, the raw material being hereafter sent to the Bureau, instead of the tabular statements, which up till then was all the bureau had for compilation. At the next decennial census, January 3rd, 1876, individual cards were used in preparing the results. It may be added that this census counted persons actually present instead of the persons domiciled in the place concerned, as was previously the case. Individual cards for marriages, births and deaths were already used in 1868. In A. N. Kiær Norway's official statistics had

over a long period a prominent and energetic leader
(1867–1913).

In *Denmark*, as mentioned above, a Statistical Bureau
entered life at New Year, 1850.   Here, as in Norway,
the domiciled population was originally counted, but
from 1870 every person who had passed the night before
in the place, whether his home were elsewhere in the
Kingdom or abroad, was enumerated.   The frequency
of marriage for each age was known since 1845, though
not separately for widowers, bachelors, etc.   As to the
births, the age of the mother was recorded since 1860
(for Copenhagen, since 1878).   Thus, somewhat early,
did *Danish* statistics on the present burning question of
the history of fertility become available.

This brief survey cannot, of course, be exhaustive, and
may easily become desultory.   At all events, it will show
that official statistics did not stand still in the period
under review, and as we shall see in a following chapter,
the theory of statistics made corresponding progress in
these decades.   Besides the organisation of a statistical
service, progress in official statistics will particularly be
found in vital statistics, as we have seen with regard to
the census.

CHAPTER XV

ECONOMIC STATISTICS IN THE CONGRESS PERIOD

64. It may seem absurd to link up the history of economic statistics with the Congress movement, for to a great extent the evolution of this branch of statistics is like a continuous stream, constantly moving on, independently of such events as the opening of the first Congress or the closing of the last one; and by choosing a definite period such as this we should arbitrarily break this continuity. If we confine ourselves to official statistical reports, it may even be justly objected that it is impossible to give a detailed systematic description of the movement, for here we often have to deal with imponderable quantities. In fact, the main result of a review of the statistical reports at various epochs will frequently consist of more or less vague impressions. Gradually the stream of facts will change its character, but it will often be impossible to describe exactly how this change took place. Comparing, for instance, the reports of 1850 with those of 1880, we cannot help being struck by the difference in quality. Undoubtedly there was great progress in this period both in reliability and completeness, but we have hardly any means of measuring the change or of valuing the increase of precision of the observations.

Still, it is not superfluous to choose certain phenomena in official economic statistics which to some extent reflect the Congress movement. It is probable that the congresses with their clear desire to promote economic statistics, to make the observations comparable and to widen the whole field, gave a real impulse to the evolution of this branch of statistics. At all events we are justified in saying that the same causes which called the congresses into being were active in the continuous progress of economic statistics.

Even where a regular statistical service had been introduced there were often great technical and administrative difficulties. It was by no means easy to find co-operators who could be entrusted with the work in the field, whether, as in *Saxony*, a statistical society was available, or whether Government officers or local authorities had to do the work. And quite naturally in many countries a suspicion lurked among the population, that the returns would be followed by higher taxes or rents. A remarkable difference can, however, often be observed in this respect. When in *Scotland* the Highland and Agricultural Society in the years 1854–7 undertook statistical investigations the farmers appeared to recognise the utility of statistics of this kind and willingly gave the information required from them, so that out of 50,000 schedules very few were unaccounted for.[1] But in *England* and *Wales*, where similar efforts began in 1866, not a few occupiers refused to make returns.[2] In *France*, too, it appeared difficult to get exact returns from the peasants. An attempt was made in 1839–40 and again in 1852, but the results do not seem to have been satisfactory; the report of the following investigation, in 1862, first speaks with more confidence of the reliability of the observations.[3] In the *United States*, where the decennial Census gives an opportunity for many-sided statistical investigations, it was openly acknowledged with regard to the manufacturing industry that the returns were often very defective; there were numerous omissions through the temporary absence of proprietors, the reluctance of

---

[1] See, for instance, Report of the Highland and Agricultural Society of Scotland to the Board of Trade, on the Agricultural Statistics of Scotland for the Year 1854. London, 1855.

[2] Major P. G. Craigie, "Statistics of Agricultural Production" (*Journ. Stat. Soc. London*, XLVI, 1883), and *Returns relating to Acreage of Land under Crops, Bare Fallow, and Grass, in the United Kingdom, in the Year 1866*. London, 1866.

[3] See Legoyt's report: *Agriculture. Résultats généraux de l'enquête décennale de 1862*, Strassbourg, 1868, pp. 47 *sq.*

some to return their production, and the utter unpreparedness of others.[1] It does not seem unjust when Mouat, speaking of the same census, maintains that the figures were heterogeneous, and the results unreliable.[2]

For the student of the economic evolution in the nineteenth century the statistical reports contain many puzzles. The well-known *German* economist Hermann, who was for many years the leader of the *Bavarian* statistics, complained of the lack of uniformity in the enumerations of live stock in various countries. There were considerable divergences regarding the date of the census; frequently the reports do not even mention at what date the enumeration took place.[3] And his successor, Georg Mayr, in a review (1874) of the various Bavarian enumerations of live stock, thinks it justifiable to assume that the census of 1810 was based on the whole on direct observations and not on estimates, even though the individual proprietors were not asked for returns.

Altogether it is interesting to see how *estimates* gradually became more or less replaced by *direct observations*. In the instructions relating to the agricultural returns for the census 1850 in the *United States* it is pointed out that in many agricultural returns the amount must be estimated, for instance, the number of bushels of wheat or oats, but under other headings, such as the number of live stock, the precise number or amount can usually be stated. But it must be acknowledged, that only if a very small part of the statistical material is found by direct observation is it possible to control the whole mass of estimates, so that the progress of reliability and exactness of the numbers is much greater than it would appear

---

[1] Ninth Census, Volume III. *The Statistics of the Wealth and Industry of the United States . . . from the Original Return of the Ninth Census, June 1, 1870 . . .* by Francis A. Walker, Superintendent of Census. Washington, 1872.

[2] Mouat, "Note on the Tenth Census of the United States of America" (*Journ. Stat. Soc. London*, XLIII, 1880).

[3] *Viehstand im Königreiche Bayern nach der Erhebung vom Monate April, 1863.* München, 1864.

to be judging from the mere proportion of direct observations. In fact, this is the underlying principle of
"sampling," and the old political arithmeticians were to
some degree justified in their efforts to draw conclusions
from the conditions in one country as to the state of
things in other countries, where only some few important
facts were known. We may safely conclude that the
period concerned made considerable progress owing to
the increasing number of direct observations.

The statisticians of those days had much to learn in
regard to *schedules*. There was a natural temptation to
crowd the schedules with questions, and it was not
always easy to give them a plain and distinct form. An
example was mentioned in a preceding chapter (art. 52).
Sometimes the reports openly acknowledge these defects.
Speaking of the industrial statistics in *Belgium*, 1846,
the Central Commission maintains—with regard to
similar investigations in 1880—that it is necessary to
leave out several occupations; by confining the investigations to certain occupations the Commission might hope
for more reliable results.[1]

A certain awkwardness in the *typographical* arrangement of the tables is often conspicuous in this period.
Frequently the material is split into so many small groups
that it is difficult to get a clear view of the facts contained
in the tables; much space was lost, many rubra being
quite empty. The *technical process* of compiling the
increasing mass of observations had also quite natural
defects in a period in which electric and other mechanical
aids were still unknown, so that the work in the statistical
offices often had the character of slow, ineffective handi-
craft. This also accounts for the decentralisation so
common in the middle of the past century. Of course,
it would be easier to get through the work by a division
of labour, the central bureau only having to draw up the
main results out of the ready-made tables sent in by the
local authorities. Unfortunately this division of labour

---

[1] C'est pour avoir trop voulu lors du précédent recensement industriel que
l'on n'a rien obtenu de sérieux (la dépêche ministerielle du 18 novembre 1880).

prevented the central bureau from an effective control of the material.

65. A few facts from the history of *agricultural* statistics may be given here.   As observed above, several countries rather early had enumerations of *live stock*. Naturally, it would appear possible, if a census of human beings could be taken, to count the cattle as well.   *Norway*, for instance, combined agricultural statistics with the census of 1835; *Denmark* counted the live stock in 1838 and repeated the census in 1861 and later every five years.   Several *German* States had similar enumerations, *Bavaria* already in 1794, *Prussia* in 1816, etc., and after the foundation of the *German Empire* a general census of the live stock was taken in 1873 and 1883. *Belgium* produced elaborate industrial and agricultural statistics in 1846, which were looked upon as a model for other countries.   In these investigations the enumeration of the live stock was included (repeated in 1856, 1866 and 1880).   In *France* enumerations of the live stock were combined with the general agricultural statistics of 1839–40 and 1852, and again in 1862 and 1882.   In this country attempts were made at finding the value of the cattle, and in spite of their defects, particularly in the case of the earlier enumerations, one may assume that the numbers obtained permitted certain important conclusions.

After the middle of the century live stock enumerations were on the whole taken in most *European* countries, at relatively short intervals.

The efforts were not confined to the live stock.   All branches of agricultural statistics were in fact taken up. The famous *Belgian* agricultural census, 1846, of which a report was published four years later, is an instance. The area under cultivation, with subdivisions, was returned; further, the quantity produced per hectare, as well as the total volume of produce, the mean weight of grain per hectolitre, the seed used per hectare, the average price of the ground, and the rent per hectare. Further, there were elaborate observations concerning the agricultural population, the labourers, wages, etc. The Commission was well aware that not all the observations were reliable, and it expresses, for instance, doubt as to the quantities produced, the numbers being possibly too low. The same system as in 1846 was used at the agricultural census of 1856, but the technical process was improved, several points being left out of the schedules where information could be otherwise obtained, for instance, as to the prices of land or wages. This simplification is one of the numerous instances of technical progress in this period.

In *France*, the first agricultural censuses appear to have been rather imperfect; but in the report for 1862 the leader of the general statistical service, Legoyt, speaks with confidence of the progressive reliability of the results. The report gives the area under cultivation, for various crops, the quantity of seed per hectare, as well as the quantity produced in the year concerned and in an average year, the prices, the value of the total produce, wages, etc. In consequence of the war of 1870 between *France* and *Germany*, no investigation for the year 1872 took place as had been planned, but the next decennial report, for 1882 (published 1887, signed by Tisserand), follows the track of the earlier enumerations. Here too we may notice marked technical progress; the schedules from each canton are, for instance, now filled in in duplicate, one set of forms being sent to the Ministry, where a collection of the original material was filed in this way. This made a revision of the results possible, and moreover enabled the central office to make fuller investigations.

In *France*, as in several other countries, *graphical* representations of statistical facts were in this period very popular; an album was published, with more or less instructive coloured graphic representations of various subjects of agricultural statistics.[1] Though works of this kind can hardly be of much scientific

---

[1] *Album de statistique agricole.* Résultats généraux de l'enquête décennale de 1882. Nancy, 1887.

importance, they may, at congresses, exhibitions and elsewhere, have contributed to arouse public interest in statistical subjects.

In *Germany* energetic efforts were made to create reliable agricultural statistics. Live stock was mentioned above. In regard to the acreage for various crops and the yearly produce, *Bavaria* seems to have taken the lead. In accordance with wishes expressed at the Congress in Paris, 1855, it was tried, with the assistance of an Agricultural Society, to find the mean annual crop produced, as well as the actual crop produced in the year concerned. These results could not be more than mere estimates, and only the average numbers were published, the other ones being suspected to be too low. But from 1863 the numbers were more reliable, a cataster having now been taken, so that the area for each crop was known. In the *German Empire* these efforts naturally were systematised, and from 1878 regular reports on the produce appeared. On the whole, most countries introduced regular statistics for area and for various crops, combined with estimates as to the quantity and value of the produce, it being possible in this way to form an idea of this important branch of agricultural economy. The progress in *Great Britain* and *Ireland* may be of interest in this respect, and a few facts concerning its evolution here may therefore be given.

In *Scotland*, the Highland and Agricultural Society undertook the task. The tables give the stock of horses, cattle, sheep and pigs, the acreage in tillage and in grass, roads, etc., and finally waste and woods. The various crops are pretty fully specified. One of the tables gives the average produce, in bushels or cwt. Here—as on the Continent—there is of course a weak point, but the reports maintain that the committees entrusted with the charge of estimating the produce have taken great pains to arrive at sound and earnest conclusions.

In *Ireland* energetic efforts were made, in order to secure statistical observations on agriculture. The observations, collected under the direction of the Government, assisted by the magistracies as well as by private persons throughout the country, go as far back as 1847. The live stock was enumerated, the value being calculated on the basis of prices in 1841; further, there was information as to the acreage of various crops, in holdings of different size, and finally estimates of production.   In 1853 an interesting investigation was made on the subject of "weeds," the area being divided into four classes; the first, areas kept generally free from weeds; the fourth, areas wholly neglected.[1]

In *England* the difficulties were much greater.   For many years estimates which were more or less reliable were the only sources of information.   Among these sources the returns by James Caird for 1850 may be mentioned.   P. G. Craigie, who was an authority on the history of this chapter of statistics, describes these returns as particularly careful estimates, made after prolonged and personal investigation, and he maintains that there is general harmony in the results.[2]   Other investigations due to private initiative deserve mention; for instance, a report for 1861, giving the result of questions to about 500 persons, as to whether the harvest was above or under the average.   But the turning-point was the year 1866, in which a foundation for agricultural statistics was laid.   From this year onwards, official reports for the whole Kingdom were published.[3]   The reports give the acreage for various kinds of crops, the live stock, although the number of horses was not included till 1869.   In 1870 a classification was made of holdings, according to their acreage, and in 1871 orchards, woods and plantations were returned.   In 1880 the holdings were distributed according to size of area in six classes and combined with the number of horses, cattle, sheep and pigs.   In 1884 the number of poultry

---

[1] *Returns of Agricultural Produce in Ireland in the year* 1854, Dublin, 1855.

[2] Major P. G. Craigie, Secretary of the General Chamber of Agriculture, "Statistics of Agricultural Production" (*Journ. Stat. Soc. London*, XLVI, 1883).

[3] *Returns relating to the Acreage of Land under Crops, Bare Fallow, and Grass in the United Kingdom, in the Year* 1866.   London, 1866.

was ascertained, though in some counties only approximately. In this year also an attempt was made at calculating the yield of cereals and other principal crops, the data being obtained through estimators specially selected for this work.

In the *United States* great efforts were made to promote agricultural statistics. The returns were combined with the decennial census and included tables concerning the produce during the year preceding the census. The same scheme of retrospective statistics was followed for *industry* and for *vital* statistics. It may of course be doubtful whether retrospective statistics of this kind can always be recommended, as omissions will easily occur. As observed above, the agricultural returns did not always rest on direct observations, estimates sometimes replacing exact numbers, but it is not possible to see from the reports to what extent this took place. The number of questions in the schedules is very large.[1] Questions are asked about acreage of land on farms ("improved" as well as "unimproved"), the "cash value" of farms, as well as implements and machinery, the live stock, with its value, also the value of animals slaughtered. There are further items about the produce during the year ending the first of June, of cereals, tobacco, wool, value of orchard products, produce of butter, cheese, hops, hemp, flax, beeswax and honey, etc. And the number of items is increased from one census to another. As a supplement to the census, yearly reports were published by the Commissioner of Agriculture, with estimates of crops, returns of cattle, etc.

66. Incomplete as the agricultural statistics must have been, still greater difficulties arose in *industrial* statistics on account of the extreme variety of establishments. Certain observations of a more elementary character may have been tolerably accurate in U.S.A. as in other countries, as, for instance, the number of establishments or the amount of labour employed, but in the

---

[1] See, for instance, *The Seventh Census of the United States*, Washington, 1853, and "Compendium of the Seventh Census," ibid., 1854.

United States the reports, just as in the agricultural statistics, aimed at much more complete information regarding such items as capital, wages, material and products, and there was a very detailed specification of establishments.

The task may have been somewhat easier in *Europe*. The frequently quoted *Belgian* Census of 1846 laid a good foundation-stone with much detailed information as to the various establishments, their machinery, etc. For the student of social history the distribution of labour according to sex, age and wages will be of interest; there are groups for children under 9 years and between 9 and 12; the wages for adult labourers range from under $\frac{1}{2}$ fr., $\frac{1}{2}$–1 fr., etc. An attempt at taking a new industrial census, in 1866, proved unsuccessful, and the results were not published. The following Census, of 1880 (the report appeared 1887), seems to have been taken under better conditions. As remarked above, the report criticises the Census of 1846, and tries to give clearer and more reliable results. The schedules have items for the persons employed, wages, numbers of working hours, value of the animal produce, etc.

In various *German* States energetic attempts at industrial statistics were made in the course of the nineteenth century; in *Prussia*, for instance, as far back as 1816. The Conference of the Customs Union expressed the desirability of statistics on this subject, and in most of the States observations on the condition of industry were collected in addition to the enumeration of the population in 1846, according to a system proposed by *Prussia*. In the following years a new enumeration was prepared (realised in 1861), but a more important census was taken in 1875 after the foundation of the *German Empire*. After this Census another was taken in 1882, upon which the subsequent enumerations were modelled. On the whole it may be said that the best impulse for statistics of this kind was given by the growing interest in *social problems* in the last decades of the century, not only in *Germany*, but all over Europe. And in this period even the difficult problem of statistics of production could be approached.

67. So far as *international trade* and *shipping* is concerned, efforts were made somewhat early, with more or less success. In *England*, for instance, regular reports on international trade began in 1834, under the direction of G. R. Porter. Whereas it must always prove difficult to get reliable information regarding the output of industrial establishments, it was possible to get from the Custom-houses comparatively good observations on *trade*, particularly imports. Not only were the quantities imported or exported of interest, but also their value. In many countries so-called *official values* were used. In *France* such average values were fixed by royal ordinance, 1826,[1] with the object of reducing imports or exports to a unity and so make the figures relating to foreign trade in the individual years comparable. In the long run, however, this method was found unpractical. In the course of time many changes took place with regard to the various articles; several goods were, for instance, originally prohibited, the import of which was afterwards permitted, and consequently new official values had to be fixed. For other articles the weights used in the tariffs were changed from nett to gross weights. A double calculation was consequently introduced, the actual prices for the year, fixed by a commission aided by the Chambers of Commerce, being added to the official values.[2] In the report for 1864 the official values disappeared altogether; for *Algiers* only they were still in use some years but given up in 1874.

Exactly the same may be noted for *England*. In a report for the years 1820–33 official rates of valuation are used, but for the last three years of this period "real or declared values" were also used. The report for 1853 says that the official value is the only value of imports that can at present be given, but as it affords no criterion

---

[1] See, for instance, *Tableau général du Commerce de la France . . . 1840,* Paris, 1841, p. vi.

[2] *Tableau général . . . 1850,* Paris, 1851, p. vii.

of the actual value of an article that valuation has not been introduced, and an endeavour is now being made at the Custom-house to obtain the calculated real value of imports.[1]

Yet an ever-increasing necessity was felt for a more exact comparison than that afforded by the actual prices of the year, and in this way the problem of *index-numbers* arose. Taking the prices at a certain moment as basis, the prices at other moments can be calculated per cent. on these numbers. Having found the movement of each individual price, the next step will be, out of the various index-numbers, to find a *total* or *general index-number* as a common expression for the increase or decrease of prices.

It may be convenient here to extend somewhat the history of this problem over the period concerned.

The problem was taken up long before the middle of the nineteenth century.[2] As far back as 1798, the English mathematician Sir George Shuckburgh Evelyn raised the question in an investigation on weights and measures, published in the *Transactions* of the Royal Society.[3] As a supplement to his investigation the author gives a table [4] containing the results of observations of prices, including wages in husbandry, reduced to index-numbers, and general index-numbers. His explanations as to method and material are however very laconic, and he even makes a curious apology for having added this table: "however I may appear to descend below the dignity of philosophy, in such œconomical researches, I trust I shall find favour, with the historian at least, and the antiquary."

Later other authors dealt with the problem, Arthur

---

[1] *Annual Statement of the Trade and Navigation of the United Kingdom with Foreign Countries and British Possessions in the Year 1853.* London, 1855.

[2] Besides the literature quoted below, I may refer to an article by Zuckerkandl, "Die statistische Bestimmung des Preisniveaus," *Handwörterbuch der Staatswissenschaften*, 3rd ed., VI, 1910, p. 1154.

[3] Sir George Shuckburgh Evelyn, *An Account of some Endeavours to ascertain a Standard of Weight and Measure.*

[4] The table was reprinted by R. Giffen in a report on index-numbers in *Bulletin de l'Institut international de statistique*, II, 1887.

Young, Joseph Lowe, George Poulett Scrope and G. R. Porter. In 1812 Arthur Young published an investigation on the value of money[1] in which he criticised Shuckburgh Evelyn's results. In order to find the movement of the price-level he uses index-numbers for various articles from which he calculates a general index-number, allotting them certain weights according to their importance, "repeating wheat five times, barley and oats twice, the produce of grass land four times, labour five times and reckoning wool, coals and iron each but once, while iron is considered the representative of all manufactures." Joseph Lowe approves of these calculations but gives no positive contribution to the solution of the problem.[2] Later, the geologist G. P. Scrope touched on the question.[3] G. R. Porter should especially be mentioned here.[4] He treated the prices of 1833–7 in the same way as Shuckburgh Evelyn, but his material was much more complete. For each month in these five years he gives the average of index-numbers for fifty articles comprising the "principal kinds of goods that enter into foreign commerce." It is his aim in this way to find "the mean variation in the aggregate of prices from month to month." Curiously enough, he did not bring those calculations up to date in the later editions of his work.

Whereas here the *arithmetic mean* was used, the famous economist W. Stanley Jevons (1835–82) in a remarkable paper (1863) proposed the *geometric mean*, calculating the mean of the logarithms instead of taking the original numbers.[5] In the case of both methods the system of

---

[1] Arthur Young, *Inquiry into the progressive Value of Money in England as marked by the Price of Agricultural Product*, London, 1812, here quoted after Joseph Lowe, *The present State of England in regard to Agriculture, Trade and Finance*, 2nd ed., London, 1823 (1st ed. appeared 1822).

[2] He is advocate of a plan which has been supported by several later economists, of agreements on wages, annuities, etc., "by changing the numerical amount in proportion to the change in its power of purchase."

[3] George Poulett Scrope, *Principles of Political Economy . . .*, London, 1833.

[4] G. R. Porter, *The Progress of the Nation*, 1838 (2nd ed., 1847, 3rd ed., 1851).

[5] W. Stanley Jevons, *A Serious Fall in the Value of Gold*, London, 1863.

allotting weights based on the plan proposed by Arthur Young can be used.

Jevons' method was of course a little more laborious than the arithmetic one, but otherwise it presents certain theoretic advantages.[1] The *arithmetic* mean, however, was generally preferred. The advocates of this method may have felt justified in employing it by the fact that the aim of these calculations was chiefly a practical one. And if all prices had a tendency to move in the same direction the arithmetic mean would naturally present itself, the various index-numbers being so to speak observations of the same quantity. This may have been a point for Porter in his calculations concerning the prices of 1833–7.

The observations now began to appear regularly. In 1859 W. Newmarch published index-numbers for nineteen articles with the New Year, 1851, as a starting-point,[2] and in the following two years he treated a similar material, extending his investigations to twenty-two articles, with the years 1845–50 as basis. From 1864 corresponding observations were published in the *Economist*, to which journal Newmarch had become a contributor. Some years later (from 1868) a general index-number was published by the *Economist*, the arithmetic mean being used. Other investigations followed the track; among these are Sauerbeck's calculations, published in the *Journal of the Statistical Society of London*.[3] Sauerbeck makes observations concerning thirty-eight commodities and he uses a very simple method of weighing, seven of the numbers being taken twice, so that altogether he bases the general index on forty-five individual numbers. In *Germany* A. Soetbeer made

---

[1] Westergaard, *Die Grundzüge der Theorie der Statistik*, Jena, 1890, pp. 218–20.

[2] W. Newmarch, "Mercantile Reports of the Character and Results of the Trade of the United Kingdom during the Year 1858" (*Journ. Stat. Soc. London*, XXII, 1859).

[3] A. Sauerbeck, "Index Numbers of Prices" (*The Economic Journal*, V, 1895). His first articles on prices appeared in 1886: "Prices of Commodities and the Precious Metals" (*Journ. Stat. Soc. London*, XLIX, 1886).

regular calculations, dealing with observations on the price-level in Hamburg; he also chose the arithmetic mean.[1]

A good foundation was laid in this way for the discussion in the following decades, in which period the material was multiplied by numerous official publications, the field being furthermore extended to *family-budgets*, etc., which served as starting-points for agreements regarding salaries and wages.

Quite naturally there was a tendency to see more in such calculations than could justly be read. A sign of this tendency may be found in a report by Neumann-Spallart in 1887, a year before his death, at the first meeting of the International Statistical Institute to the foundation of which he had essentially contributed.[2] He proposed to calculate a great many index-numbers to illustrate the economic, social and moral condition of various countries. He intended to make an investigation on this subject, but his early death prevented the realisation of his plan. It is not improbable that an attempt to deal with these heterogeneous quantities would have been a disappointment for him.

Naturally various other indexes came into use in this period, in order to facilitate comparisons. In *shipping statistics*, for instance, observations on the freight carried by the various categories of ships was used to find the effectiveness of the tonnage. One ton in a steamer was considered equal to three tons in a sailing-vessel, although the proportion four to one was sometimes recommended.[3] *In agricultural statistics* the *live-stock* was compared by

---

[1] A. Soetbeer, *Materialien zur Erläuterung und Beurteilung der wirtschaftlichen Edelmetallverhältnisse und der Währungsfrage*, Berlin, 1885.

[2] Rapport de M. de Neumann-Spallart, *Mesure des variations de l'état économique et social des peuples* (Bull. de l'Institut international de statistique, II, 1887).

[3] *Tabeller vedkommende Norges Skibsfart i Aaret*, 1871 Christiania, 1873, p. viii.

A. N. Kiær, *Mouvement de la navigation, Statistique internationale*, Christiania, 1892, p. xxxiv.

reducing all numbers to cattle, making, for instance, one horse equal to $1\frac{1}{2}$ head of cattle.[1]

This method presented, of course, a short-cut in comparison with the *French* method of finding the *value* of the various species; but it was not possible to obtain absolute uniformity.[2] In our days those indexes are to a great extent growing obsolete.

A curious attempt was made by C. L. Madsen to find a mathematical law for the number of telegrams in the international traffic between two countries. His idea was to calculate this number from the value of the trade between two countries and from the tonnage of the ships which carried the goods.[3]

The present author tried to show the defects of the formula, one objection being the very trivial one that the number of telegrams would change, if the unity was altered, for instance, from £ to francs.[4]

68. It is a sign of the growing confidence in statistical material during this period that energetic attempts were now made at getting a bird's-eye view of the economic conditions of all countries in the civilised world. Official statistical reports on international trade were abundantly at hand, and so far as the production of several commodities is concerned a growing amount of material was accessible through the help of Chambers of Commerce and other institutions, brokers and other private firms, etc. An increasing number of journals took an interest in the collection of observations, such

---

[1] See, for instance, *Tabeller over Kreaturholdet i Danmark den 16 Juli, 1866*, Copenhagen, 1868, p. ii.
Maurice Block, *Puissance comparée des divers états de l'Europe*, Gotha, 1862, p. 115.

[2] *Kreaturholdet den 15 Juli, 1893*, Copenhagen, 1894, p. xiv.

[3] C. L. Madsen, *Den sandsynlige Lov for den internationale Telegraftrafik*, 1876; *Recherches sur la loi du mouvement télégraphique international*, 1877. The formula was $T = C(\sqrt{V.N} + N_1 + N_2)$, C being a quantity depending on various circumstances, V the value in £, N the number of registered tons, N_1 and N_2 representing the tons of ships belonging to one of the countries and carrying goods between the other country and foreign ports.

[4] Westergaard, "Den sandsynlige Lov for den internationale Telegraftrafik," *Nationaløkonomisk Tidsskrift*, 1878. In the following number of the journal C. L. Madsen replied, maintaining the correctness of his formula.

as the *Economist*, the *Statist*, etc.   One of the promoters
of this branch of statistics was Fr. X. Neumann-Spallart,
who has frequently been quoted above.   From estimates
and direct observations he tried to get a more or less
reliable picture of the production and consumption all
over the world of certain commodities, his first attempt
of this kind being a separately printed contribution to
Behm's geographical year-book, 1874.[1]   He deals here
with the production of gold and silver, of coal and iron,
coffee and cotton; further, he treats such subjects as
railways, shipping and telegraphs, also international
trade, giving the value of exports and imports from
1866–73.   He is well aware of the defects in the material,
regretting that many of the numbers are only approximate.   But the successive publications [2] grow richer and
more reliable, and he extends the field more and more,
for instance, to the production of, and trade in, cereals.
Even though these observations had obvious defects, as
for instance with the well-known difference between export and import values, they enabled him to draw interesting conclusions about the world-market with its
numerous circles in the interchange between the individual countries, following, as he did, the origin and the
sale of the commodities concerned.   On the whole, these
surveys testified favourably to the progress of economic
statistics in the congress period.

[1] Fr. X. Neumann, *Uebersichten über Production, Welthandel und Verkehrs-
mittel*, Wien, 1874.

[2] Fr. X. Neumann-Spallart, *Uebersichten über Produktion, Verkehr und
Handel in der Volkswirthschaft*, Jahrgang 1878, Stuttgart, 1878.   Later followed four *Uebersichten der Weltwirthschaft*.   Juraschek continued the work
for the years from 1885.

CHAPTER XVI

PROGRESS OF VITAL STATISTICS AND OF THEORY IN THE CONGRESS PERIOD

69. To combine in this chapter the progress of vital statistics with that of theory may sound strange, but still it is not arbitrary as there is in fact a close connection between these subjects. Theoretical problems were to a great extent engendered in vital statistics. It is impossible to describe the evolution of vital statistics without entering upon theory.

It has been shown that several of these problems had already found their solution before the middle of the past century. Various rational life-tables are evidence of the progress. It was now possible to get reliable results in questions hitherto abandoned to instinct and guess-work. Yet much was still needed, the progress being anything but uniform. By looking through the statistical literature of the Congress Period we cannot help being struck by the large number of authors who were still using obsolete methods of investigation. This, for instance, was so in the case of *medical* statistics. Here, for many years after the middle of the century, a great confusion ruled: medical men interested in statistical observations frequently had not learnt to master the relatively few theorems of statistics needed in this field. Other authors, who undertook the task of *popularising* statistical results, and who thereby promoted the public interest in statistics, were often themselves anything but clear with regard to the more intricate prcblems. And the growing army of *practising* statisticians could not always boast of a good theoretic equipment. Much would be reached if only the old-fashioned ineffective weapons could be replaced by a modern equipment.

The technical progress in collecting statistical observations has already been mentioned. Undoubtedly census results grow more and more reliable in the first decades after the middle of the century. Estimates became a mere exception, and if they are found unavoidable they are now much more easily handled than before.[1] It is true that the results were not unanimously accepted. The *American* census results, for instance, had evident defects, openly acknowledged in the public reports; such defects were particularly found in the retrospective mortality statistics which, after 1850, were linked to the census. And in England William Farr was called upon to defend the results of official vital statistics against criticism, a task which he performed with his usual energy.[2]

At all events there seems to have been a growing confidence in the general results, and, just as in *economic* statistics, the great increase in the mass of statistical observations made it natural to attempt a *compilation* and a comparison of the results from various countries. As in the age of political arithmetic—but with more success—attempts would be made at finding the total population of *Europe* or even of the whole world. At first, however, the difficulties were great. Brachelli in the first edition of his statistics of European States (1853)[3] did not venture to give a grand total for the European population. Nor did Maurice Block make

---

[1] As an instance can be mentioned the *Bengal* census, 1872, where in certain districts a detailed enumeration was excluded: Beverley, "The Census of Bengal" (*Journ. Stat. Soc. London*, XXXVII, 1874).

[2] W. Lucas Sargant: (*a*) "On Certain Results and Defects of the Reports of the Registrar-General," (*b*) "Inconsistencies of the English Census of 1861, with the Registrar-General's Reports; and Deficiencies in the Local Registry of Births" (*Journ. Stat. Soc. London*, XXVII–XXVIII, 1864–5). W. Farr, "On Infant Mortality and on Alleged Inaccuracies of the Census," ibid., 1865. See also T. A. Welton, "On the Classification of the People by Occupations; and on other Subjects connected with Population Statistics of England," ibid., XXXII, 1869.

[3] Brachelli, *Die Staaten Europas*, Brünn, 1853.

any attempt of this kind in a corresponding work.[1] More courageous was Fr. Kolb in the first edition (1857) of his handbook of comparative statistics.[2] Later in the congress period official statistics made such progress that compilation could be much more easily undertaken. Weighty contributions to this branch of international statistics were due to Bodio, who was for many years the leader of Italian statistics.[3] The next step would be, by means of corresponding observations on the *movements* of the population of the world, to give an account of the increase of population through excess of births over deaths and to survey the migrations from one part of the globe to another, in order to get the horoscope of the population. This, however, had to be left to the subsequent period.

70. As remarked above, *medical* statistics were much behind in the middle of the past century, and for many years after that epoch antiquated methods still prevailed.

As had been the case before, statistical dilettanti were tempted to draw definite conclusions from a one-sided material, as, for instance, death-lists. As Lombard had done in treating observations for Geneva (art. 55), and as Guy continued during his long life (he died in 1885) with regard to English observations, so did W. C. de Neufville and Lübstorff, the former dealing with observations from Frankfurt-a-M.,[4] the latter with material from Lübeck.[5] The result was a hopeless confusion. With regard to duration of life, for instance, de Neufville assigns a favourable position to bakers as compared with shoemakers (51·5 and 47·3 years respectively), whereas Lübstorff arrives at the opposite result (48·4 and 53·2 years). And whilst the latter finds the same

---

[1] Maurice Block, *Puissance comparée des divers états de l'Europe*, Gotha, 1862.

[2] G. Fr. Kolb, *Handbuch der vergleichenden Statistik*, Zürich, 1857.

[3] *Movimento dello stato civile anni 1862–78*. Introduzione con confronti di statistica internazionale, Roma, 1880.

[4] W. C. de Neufville, *Lebensdauer und Todesursachen 22 verschiedener Stände und Gewerbe*, Frankfurt-a-M., 1855.

[5] H. Lübstorff, *Beiträge zur Kenntniss des öffentlichen Gesundheitszustandes der Stadt Lübeck*, Lübeck, 1862.

duration of life for the medical profession and the clergy, de Neufville maintains that there is a very great difference in favour of the clergy. The confusion was made still greater by the investigations of Escherich, who follows the system of comparing the age distribution of *living* persons in various professions instead of using the death-lists as basis. He finds that the "liberal" professions have unfavourable health conditions, an assertion which is not confirmed by later investigations.[1]   In a frequently quoted compendium of medical statistics published in 1865, F. Oesterlen gave a survey of investigations of this kind, without making any serious attempt to explain the differences.[2]   The mortality in Paris was treated by Trébuchet in various articles. He tries to find the mortality in several professions, basing his calculations on death-lists and on the census 1851, but without taking the ages of the living into consideration.   Little could be learnt in that way.[3]

If authors of medical statistics were unable to get reliable results as to the duration of life in various classes of society on the basis of one-sided observations on deaths only, it was tempting to follow another track in order to measure the effect of economic conditions, habits of intemperance, occupation, etc., the method simply consisting in the calculation of the *relative frequency* of certain diseases or causes of death according to the experience in hospitals or in a whole population. To this category belongs an investigation by Hannover on the diseases of artisans in Copenhagen,[4] and in the above-quoted report of 1855 de Neufville treats the distribution of deaths, finding, for instance, that out of 100 deaths among tailors 43 were due to tuberculosis

[1] Escherich, *Hygienisch-statistische Studien über die Lebensdauer in verschiedenen Ständen*, Würzburg, 1854.

[2] Fr. Oesterlen, *Handbuch der medicinischen Statistik*, Tübingen, 1865.

[3] Trébuchet, "Recherches de la mortalité dans la ville de Paris," Années 1851, 1852, 1853 (*Annales d'Hygiène publique*, 50, 1853, 2me série, 7, 1857; 9, 1858).

[4] Hannover, *Haandvarkernes Sygdomme*, Copenhagen, 1861; see also *Annales d'Hygiène publique*, 2me série, 17, 1862.

and 7 to diseases of the nervous system, whereas the corresponding numbers for merchants were 24 and 17 per cent. According to an English investigation, the relative mortality from zymotic diseases in streets with a well-to-do population was 18 per cent., whereas the corresponding proportion in poorer streets was 31.[1] It must be acknowledged that observations of this kind may be useful as symptoms if, for instance, the difference is enormously great, as may be the case with the mortality from small-pox among vaccinated and non-vaccinated persons. But in most cases such investigations will obviously be irrelevant.

The influence of occupation on health is of course only one chapter of medical statistics. There are several other subjects of interest. Here, too, the authors were generally unable to draw safe conclusions. Oesterlen is aware of Poisson's contributions to the calculus of probability and of the test which Gavarret recommended.[2] But he objects that Gavarret's method is neither so accurate nor so indispensable as asserted (l.c., p. 62), and he seems to prefer the elementary method of dividing the material into various groups, all of which would show identical results if the number of observations is sufficiently large. Most authors took no notice at all of tests of this kind.[3]

Several problems presented themselves in the reports of hospitals; for instance, the chance of dying from an attack of a certain disease—the *lethality*, or the duration of the treatment in the hospitals.

Isolated investigations were sometimes added as to the duration of certain diseases, as, for example, cancer or phthisis, dating the calculations from the moment when the first attack was observed.[4] Neison tried to utilise the experience of the oldest German life-office

---

[1] Oesterlen, l.c., p. 894.

[2] Gavarret, *Principes généraux de la statistique médicale*, Paris, 1840 (see above, art. 54).

[3] An exception was J. Hirschberg: *Die mathematischen Grundlagen der medizinischen Statistik elementar dargestellt*, Leipzig, 1874.

[4] Oesterlen, l.c., pp. 377, 431.

(the Gotha society) regarding causes of death.[1]   The deaths were classified according to cause and age, and in addition the average duration of the assurance was calculated.   The material was, however, rather limited and the principle followed somewhat arbitrary, so that this experiment could hardly be considered successful.

One fertile subject was the description of *epidemics*, as, for instance, cholera.   Here William Farr gave weighty contributions, in a time when bacteriology was as yet unknown.   He localised the cholera epidemic in London, 1866, essentially to the field of one of the companies which supplied the population with water.[2]   His biographer states that his address at the congress in Florence the following year, describing the sudden outbreak of the epidemic "and its as sudden cessation when the supply of polluted water which was its cause ceased, was listened to with breathless attention" (l.c., p. xxii).

In order to get really incontestable observations on the influence of occupation on health, it was necessary to get complete observations on deaths as well as on persons exposed to risk.   Here two roads presented themselves.   On the one hand it often proved possible to get trustworthy material of this kind for certain classes of society; on the other, the use of census material and official lists of deaths might be tried with due attention to profession and age distribution.   In regard to the first alternative, several monographs were published in the course of time, based on correct principles.   As an instance, an investigation by Bailey and Day on the British peerage may be quoted,[3] Ch. Ansell, jun., treated the upper and professional classes of society by means of observations collected by a life-office.[4]   Other authors, such as Hodgson and Stüssi, dealt with the mortality

---

[1] F. G. P. Neison, *Vital Statistics*, 3rd ed., 1857, pp. 172 *sq.*

[2] Memorial Volume, pp. 369 *sq.*

[3] Bailey and Day, "On the Rate of Mortality amongst the Families of the British Peerage" (*Journ. of the Institute of Actuaries*, IX, 1861); (see also *Journ. Stat. Soc. London*, XXVI, 1863).

[4] Ch. Ansell, *Statistics of Families in the Upper and Professional Classes*, National Life Assurance Society, London, 1874.

of the clergy,[1] or F. G. P. Neison[2] and Gussmann[3] with the medical profession. Retired civil officers formed another example treated by Achard.[4] Life offices, widow funds and friendly societies were often able to give material for investigations of this kind.[5] This was the case with a paper by J. Stott dealing with the health of innkeepers and publicans.[6]

Miners were several times the subject of investigations, for instance by G. Zeuner,[7] whose contributions related to miners in Saxony (1853–4), whereas William Farr wrote about miners in Great Britain.[8] Railway-servants formed another social class for which it was possible to get reliable observations.[9]

These contributions added much to the knowledge as to the mortality in various occupations, but it was of course a drawback that the periods of observations varied, and that other circumstances were frequently very different. It would hardly be possible to eliminate all disturbing causes, so that occupation as the chief cause could be isolated. In this direction, however,

[1] Hodgson, *Observations in Reference to Duration of Life amongst the Clergy of England and Wales*, with a Supplement by Samuel Brown, London, 1865. H. Stüssi, "On the Mortality of the Clergy. Translated and abridged by Bumsted" (*Journ. Institute of Actuaries*, XVIII, 1874).

[2] F. G. P. Neison, *Vital Statistics*, 3rd ed., 1857 (on the Rate of Mortality in the Medical Profession).

[3] Gussmann, *Statistische Untersuchungen über die Mortalitätsverhältnisse im ärztlichen Stande*, Tübingen, 1865.

[4] Achard, "Recherches statistiques sur la longévité des pensionaires civils de l'état" (*Journal des Actuaires Français*, VIII, 1879).

[5] For instance, Broch, "Dødeligheden blandt Indskyderne i den norske Enkekasse," *Nyt Magazin for Naturvidenskaberne*, 21 Christiania, 1875.

[6] J. Stott, "On the Mortality among Innkeepers, Publicans and other Persons engaged in the Sale of Intoxicating Liquors—being the Experience of the Scottish Amicable Life Assurance Society during Fifty Years 1826–76" (*Journ. Inst. Act.*, XX, 1876).

[7] See an abstract in *Zeitschrift des Kgl. sächsischen statistischen Bur.*, IX, 1863.

[8] Appendix B to Report of the Commissioners appointed to inquire into the Condition of all Mines in Great Britain, 1864. For further literature, see Westergaard, *Die Lehre von der Mortalität*, 2nd ed., Jena, 1901, pp. 579 *sq.*

[9] G. Behm, *Statistik der Mortalitäts- Invaliditäts- und Morbilitätsverhältnisse bei dem Beamtenpersonal der Deutschen Eisenbahn-Verwaltungen*, Berlin, 1876–85.

English official statistics gave weighty contributions, particularly through William Farr, and it is not too much to say that the backbone of all investigations in this field is to be found here.   In this respect the Census of 1851 [1] was important.   For a large number of occupations the persons were counted, and out of this material twelve occupations were selected, for which the age distribution was found and compared with the corresponding cases of death in the census year, so that correct rates of mortality could be calculated.   In this way a large area was opened up for exploration.   At future censuses the number of selected occupations could be increased, so that gradually a large amount of material could be available.   It might be objected against the experiment of 1851 that the observations on deaths only covered one calendar year.   As a matter of fact, the year 1851 was relatively healthy and the observations did not therefore give a true picture of the average mortality at that time.   It might, however, be supposed that the favourable condition of health in the census year would prove effective in all the twelve selected occupations, so that a comparison of their relative health would not be excluded.   This was proved by later investigations.[2]   By the next census (1861) the deaths for the two years 1860–1 were used.   The next census again, in 1871, gave material for corresponding statistics, based on deaths for the census year.   The inner harmony of many of the results testify to the reliability of this material.[3]   Further progress was made, particularly as regards the *causes of death*.   William Farr worked energetically to improve the nomenclature and to secure

---

[1] Fourteenth Annual Report of the Registrar-General of Births, Deaths, and Marriages in England, London, 1855.

[2] Supplement to the Twenty-Fifth Annual Report of the Registrar-General . . . London, 1864, and Supplement to the Thirty-Fifth Annual Report—1875.   As to Farr's part in the work, see the above-quoted Memorial Volume by Noel A. Humphreys, London, 1885.

[3] An attempt to give a comprehensive review of these results was made by the present author in *Die Lehre von der Mortalität und Morbilität*, 1st ed., Jena, 1882.

a regular distribution of the deaths according to age.[1]
In this way it would be possible to calculate rates of
mortality from various causes of death for the *whole*
population, and the next step would be to provide corresponding material for various selected *occupations*, thereby
finding an explanation of the great differences in mortality in different occupations. This was done by
William Farr's successor, William Ogle, for a considerable number of occupations. The numbers of
deaths in 1880–2 from various causes were found by
sampling, dividing the total mortality in the industry
concerned by the proportion existing in the sample.[2]
Later, further progress was made, the general result
giving a much greater uniformity than at first expected,
so that a comparatively small number of influences will
generally explain the main difference observed (accidents,
dust-inhalation, habits of intemperance, etc.).

71. Returning to the subject of life-tables, we are
able to register several important contributions from
life-offices and friendly societies. A. G. Finlaison (son
of J. Finlaison) made a report (1860) on Government
Annuities.[3] Twenty English life-offices co-operated to
provide a mortality-table on the basis of their experience. The introduction to the report was written by
Samuel Brown.[4] Various American life-tables were also
published in this period and the above-mentioned German life-office, Die Lebensversicherungsbank zu Gotha,
founded in 1829, published a jubilee report in 1880,
containing fifty years' experience.[5] A few years later
(1883) a co-operative investigation followed, on the ex-

---

[1] Memorial Volume, pp. 314 *sq.*

[2] Supplement to the Forty-Fifth Annual Report of the Registrar-General
of Births, Deaths, and Marriages in England. London, 1885, see Introduction, p. xxviii.

[3] A. G. Finlaison, *Tontines and Life Annuities.* Report and Observations
on the Mortality of the Government Life Annuities, London, 1860.

[4] *The Mortality Experience of Life Assurance Companies.* Collected by the
Institute of Actuaries, London, 1869.

[5] A. Emminghaus, *Mittheilungen aus der Geschäfts- und Sterblichkeits-
statistik der Lebensversicherungsbank für Deutschland zu Gotha für die fünfzig
Jahre von 1829 bis 1878.* Weimar, 1880.

perience of twenty-three *German* life-offices; the tables were graduated by Zillmer.[1]

In this epoch the representatives of official statistics were by no means always clear with regard to the problems in vital statistics.   In Prussia, for instance, Dieterici, who in 1844 followed Hoffmann as leader of the official statistics, published in the transactions of the Royal Academy in Berlin various contributions, which bear out this lack of clearness.   In one of these contributions[2] he tries to find the mean duration of life.   He mentions various methods, as for instance that of calculating the mean age of the living population, but he seems to prefer the average age at death, and his successor, Engel, did not go beyond this standpoint.[3]   Engel maintains that on the whole there is not a very great difference between the mean age of the *living* at the moment of a census, and the mean age at *death* during a long period of years, and he proceeds to treat the deaths in Prussia, 1816–60, from this point of view. It is superfluous to explain the logical mistake which he made.   In *Belgium* Heuschling made a curious attempt to calculate a life-table on deaths only[4] under the supposition that the number of births exceeded that of the deaths by about one-fourth, calculating, however, the rate of mortality in the first year of life separately, from the births and the deaths.   After the first year of life he simply added one-fourth to all the deaths, and he felt justified in calculating a life-table from this material alone.   A discussion in the *Journal des Économistes* arose (1854–5) in which Liagre and Quetelet took part, trying to make Heuschling see that his method gave precisely the same result, as if he had taken the deaths,

---

[1] *Deutsche Sterblichkeitstafeln, aus den Erfahrungen von dreiundzwanzig Lebensversicherungs-Gesellschaften,* Berlin, 1883.

[2] Dieterici, *Ueber den Begriff der mittleren Lebensdauer und deren Berechnung für den preussischen Staat, Akademie der Wissenschaften zu Berlin, 1858.* Berlin, 1859.

[3] Engel, "Die Sterblichkeit und die Lebenserwartung im preussischen Staate und besonders in Berlin," *Zeitschr. des Kgl. Preussischen Stat. Bureaus,* I–II, 1861–2.

[4] Westergaard, l.c., 2nd ed., 1901, pp. 99 *sq.*

without any addition, as basis. Heuschling appears to have been misled by the separate calculation of infant-mortality, the other rates consequently being slightly different from the rates calculated on the basis of the deaths through the whole life.

Other statisticians were aware of the conditions necessary for a rational life-table, but they saw a difficulty in the defects of the census. As Euler had proposed in the eighteenth century, they took the growth of the population into consideration. Liagre calculated a life-table for *Belgium* based on the death-lists, under the supposition that there was a regular yearly *geometric* increase of 6·2 per mille. In *France* a similar calculation was tried by Guillard under the supposition of a yearly *arithmetical* increase.

Here may be mentioned the interesting attempts in *Bavaria* at the construction of life-tables from births and deaths, using observations on emigration and conscription as a means of control. Hermann, who initiated this experiment, proposed to follow a generation from the cradle to the grave.[1] This, of course, is changing the problem, for in this way the result arrived at would represent the rates of mortality of the *past* instead of those of the *present* time. By that time it was already evident—though not accepted by all statisticians—that great changes in health were taking place. Moreover, it was a drawback that the material was incomplete: the age distribution of the deaths being known only since 1834–5 and birth statistics being only available from 1817–18. It was therefore impossible to follow any generation further than to the age of 42. Hermann does not describe his method in detail, though its principle is quite clear: if the number of deaths in a given age as well as the number of births in the corresponding period is known, it is possible to calculate how many persons among 10,000 new-born children would die at that age, and out of several quotients of this

---

[1] *Beiträge zur Statistik des Königreichs Bayern*, Heft III (1854), Heft VIII (1857) and Heft IX (1859).

kind average results could be found which will easily lead to the construction of a life-table.

Even though much doubt still prevailed as to the correctness of the *censuses* and though the ages distribution was in many cases unsatisfactory (as for instance in *Prussia*), the census was acknowledged as a good basis on which to work, and a number of life-tables appeared, calculated from enumerations of the populations and the death-lists in a corresponding period. Quetelet calculated a table from the results of the *Belgian* census of 1846 and the deaths 1841–50.[1] At about the same time E. Horn published a table for *Belgium*, on a somewhat similar basis, taking in also the conjugal condition.[2] A similar table was calculated for *Holland*. Further, the *Norwegian* sociologist and statistician, Eilert Sundt, deserves mention. As basis for his life-table for *Norway* he used the three censuses of 1825, 1835 and 1845, and the death-lists 1821–50.[3] He threw much light on the cyclical movements of the population: an increase in the number of weddings causing an increase of births; and this again being followed by a fresh increase in marriages many years after.[4] In England a life-table III was constructed by William Farr from the censuses 1841 and 1851 and from the deaths in the seventeen years 1838–54.[5] He hoped that this table would find general application in life insurance, but quite naturally the life-offices preferred their own experience.

The statistics of frequence and duration of *sickness* derived from the experience of *friendly* societies made conspicuous progress in the period concerned, particularly in *England*. Neison's contributions in an earlier period have been mentioned (art. 56). They were

---

[1] Quetelet, *Nouvelles tables de mortalité pour la Belgique* (Bull. de la commission centrale de statistique IV, 1851; see above, art. 57).

[2] *Journal des Économistes*, 2me série, IV, 1854.

[3] Eilert Sundt, *Om Dødeligheden i Norge*, Christiania, 1855.

[4] Eilert Sundt, *Om Giftermaal i Norge*, Christiania, 1855.

[5] *English Life-Table, Tables of Lifetimes, Annuities and Premiums*, with an Introduction by William Farr, London, 1864.

supplemented by an official report, written by J. Finlaison.[1]   The large affiliated order of friendly societies, "the Manchester Unity of the Independent Order of Odd Fellows," published three reports in this period, written by Henry Ratcliffe; they have their weakness, but on the whole they added considerably to the subject.[2]   From other countries came interesting contributions; for instance, in *France*, that of Hubbard.[3]   In *Germany* the experience of an assurance company, Gegenseitigkeit, was treated by Heym,[4] and in *Italy* at about the same time a report appeared on the statistics of several mutual aid societies.[5]

Through these investigations it was possible to judge with greater accuracy than before of the economy of friendly societies, although several problems, such as that of protracted disease, still required more complete observations.

The material of friendly societies helped to supply information with reference to the influence of occupation on health.   Neison had already tried to compare the health among persons working "indoors or outdoors," and whose occupation required little or great exercise.   A great drawback with regard to such investigations was the lack of uniformity among the societies.   It is easier to obtain clear statistical results from life insurance than from sickness insurance, each of the societies having its separate practice and rules. J. Finlaison took up the same problem in his report. Neison had also given particulars for a number of occu-

---

[1] J. Finlaison, *Friendly Societies: Sickness and Mortality*.   Ordered by the House of Commons to be printed 16 August, 1853, and 12 August, 1854.

[2] Henry Ratcliffe, *Observations on the Rate of Mortality and Sickness existing among Friendly Societies. Particularised for Various Trades, Occupations and Localities*.   Manchester, 1850.   The Second Report: Colchester, 1862. Finally: Independent Order of Odd-Fellows, Manchester Unity Friendly Society.   Supplementary Report, July 1, 1872.

[3] Hubbard, *De l'organisation des sociétés de prévoyance*, Paris, 1852.

[4] K. Heym, *Anzahl und Dauer der Krankheiten in gemischter Bevölkerung*, Leipzig, 1878.

[5] *Statistica della morbosità . . . delle malattie presso i soci delle società di mutuo soccorso*, Roma, 1879.

pations and the same was done by Ratcliffe in his first two reports.   An *Italian* report may also be noted here. It was, however, hardly possible to arrive at perfectly safe conclusions.   Ratcliffe's material was, for instance, frequently too small; in his first reports he gives a life-table for coopers on the basis of 25 deaths and 2,201 years of life only.

72.   One of the most attractive features of this epoch is the energy with which *theoretic problems* in vital statistics were attacked.   Here *German* statisticians were the pioneers, and among these especially Knapp, Zeuner and Lexis.   The first of the contributions concerned was published by Knapp (1868).[1]   By that time Zeuner had for many years been engaged in problems of life insurance and vital statistics.   His first impulse after having read Knapp's treatise was to lay his own investigations aside as superfluous, but fortunately he changed this resolution, and in a standard work he gave Knapp's results an elegant mathematical form.   These two publications are therefore closely connected.[2]

Zeuner describes the movements of a population by means of a system of three co-ordinates.   On one of the axes the moment of birth can be recorded, on another the present age of the persons concerned, on the third, the numbers of survivors are measured.   The movements of the population will be represented by a surface, the surviving at various ages of a given generation by a curve on a plane cutting the surface, and a series of corresponding curves in parallel planes will give a picture of the decreases of successive generations. Another curve on the surface will represent the numbers of persons who are present at a certain moment, for instance, at a census, etc.

---

[1] G. F. Knapp, *Ueber die Ermittlung der Sterblichkeit aus den Aufzeichnungen der Bevölkerungs-Statistik*, Leipzig, 1868.   Later appeared, by the same author, *Die Sterblichkeit in Sachsen nach amtlichen Quellen dargestellt*, Leipzig, 1869, and *Theorie des Bevölkerungs-Wechsels.   Abhandlungen zur angewandten Mathematik*, Braunschweig, 1874.

[2] Gustav Zeuner, *Abhandlungen aus der mathematischen Statistik*, Leipzig, 1869.

In his treatise Knapp discerns various groups ("Hauptgesammtheiten") of deaths, *first*, persons belonging to the same generation, who died at a given age, *second*, persons of the same generation, dying within a certain space of time, and *finally* persons who died at a given age in a certain space of time. By Zeuner's stereometric representation it is easy to ascertain these groups. Knapp's definition of the "Hauptgesammtheiten" leads back to the well-known Halley problem concerning infant mortality. William Farr, among others, would solve this problem approximately by means of interpolation. Knapp proposed to deal with it by means of direct observation, the deaths in a calendar year being recorded not only as to age, but also as to birth-year.

In his treatise, 1869, on the mortality in *Saxony*, he enters practically upon the question, basing on experience from the Duchy of *Anhalt*, 1860–6. This calculation was therefore frequently called the *Anhalt* method, though at about the same time K. Becker had applied it in *Oldenburg*. In *Holland* the method was already proposed in 1866, by van Pesch, independently of the German authors.[1]

The advantage of the graphic representation advocated by Knapp and Zeuner, whether they made use of three dimensions, as did the latter, or only of two, as Knapp has done, was not confined to these practical results. In fact, it was easy to solve many problems concerning displacements within a population, particularly by the help of the continuous method, studying the movements in an infinitely small moment of time. In this way simple formulas of approximation could be derived. But it goes without saying that the *geometric* analysis can be easily replaced by *algebraic* methods, even though the former in many cases will be found more attractive.

The two treatises which I have quoted here do not stand alone. As a worthy supplement a contribution

---

[1] Van Pesch afterwards constructed the official life-table for Holland: *Sterftetafels voor Nederland.* Haarlem, 1885.

by W. Lexis [1] can be mentioned. He too uses geometric methods. Nor were investigations of this kind confined to the Continent. Thus in 1875 Verwey published a paper [2] the programme of which was to find and express in function of time and space the laws which regulate the quantitative and qualitative changes which populations undergo. Without apparently knowing the *German* contributions he applies similar methods (with two dimensions) and continuous formulas.

On the whole, the prejudice against continuous calculations had to a great extent disappeared in this period. A remarkable contribution to actuarial calculations was made by W. S. B. Woolhouse, [3] who applied the principle with great advantage on several of the chief actuarial problems.

The scientific spirit in Knapp's and Zeuner's investigations manifested itself in several contributions to vital statistics which saw light in the following years. Among *German* statisticians Becker and Böckh can be mentioned.

The former wrote a report in 1874 on the calculation of life-tables [4]; the first life-table for *Germany* was calculated under his auspices in his capacity of the leader of the official statistics of the Empire. R. Böckh, who was chief of the municipal statistical office in *Berlin* from 1875, had in view the problem of establishing a regular account of the population under due regard to migrations and to the displacements between the groups of legitimate and illegitimate children.

The refined methods signified important *scientific* progress. A comparison of the *practical* results with those of the more primitive calculations will show that as a rule the difference was not very great. This is a very

---

[1] W. Lexis, *Einleitung in die Theorie der Bevölkerungsstatistik.* Strassburg, 1875.

[2] A. J. Verwey, "Principles of Vital Statistics" (*Journ. Stat. Soc. London,* XXXVIII, 1875).

[3] W. S. B. Woolhouse, "On an Improved Theory of Annuities and Insurances" (*Journ. Inst. Act.,* XV, 1870).

[4] K. Becker, *Zur Berechnung von Sterbetafeln an die Bevölkerungsstatistik zu stellende Anforderungen,* Berlin, 1874.

valuable experience.   It will evidently add to the general confidence in official statistics that these refinements resulted in small corrections only and that even rough methods could lead to tolerably reliable tables of mortality.

However, these investigations had a wider range.   In fact, the horizon was now much widened.   Without changing principles the method could be applied to all problems concerning the movements of a population. This is the case with *marriage*-statistics, and it also applies to *birth*-rates according to age—both for married and unmarried women.   Here, as mentioned previously, *Scandinavian* statistics gave valuable contributions.   And in close connection with life-tables, observations on *disease* and *invalidism* could be treated from a similar point of view.

The problem of *migrations* (within the borders of a country, or from one country to another) is of similar nature, even though somewhat more complicated.   And if the material were at hand, analogous studies in *criminal* statistics could be made, based on individual observations, leading to a clearer insight into the life of criminals, their relapses, etc., than Quetelet had been able to give.

This will all help to explain the enormous production in vital statistics in the last decades of the nineteenth century.

73. Closely related to this subject was the evolution of *anthropometry*, which had already found a scientific basis before the middle of the century, in so far as Quetelet and others had proved that the distribution of certain measurements of the human body followed the binomial law, so that the theorems of the calculus of probabilities could be applied here.   What remained to be done was a critical revision of existing material supplemented with fresh collections of correctly made observations.   The handbooks of this period testify to the efforts in this direction.[1]   Various anthropological so-

---

[1] See, for instance, Topinard, *Éléments d'Anthropologie générale*, Paris, 1885; Quetelet (1) "Physique sociale," ibid., 1869; (2) "Anthropométrie," ibid., 1871.

cieties (as in France, founded by Broca and others in 1859) promoted public interest in the problems. In the *United States* a larger volume of detailed material was collected during the civil war, and treated by J. H. Baxter[1] and B. A. Gould.[2] *Italy* provided a large amount of material and several statisticians, for instance Perozzo, treated the observations.[3]

A brilliant practical application of anthropometric observations was initiated by Alph. Bertillon (1881).[4] The police in Paris, and subsequently in many other parts of the world, accepted his proposals, which greatly facilitated the *identification* of criminals. The system consisted simply in the registration of various measures, as, for instance, dimensions of the head. If these measures were arranged according to size, it was possible to form so many groups that only very few persons would belong to the same group. The basis of this system is evidently that the measurements of the various parts of an individual body are not in proportion to one another, so that a certain dimension might be calculated from another measurement; a tall man may have a short foot, etc.; in other words, the distribution around the mean of a given measurement is relatively independent of other distributions. The technical improvement of the system during the following fifty years (especially with regard to finger-prints) forms an interesting chapter of Anthropometry.

Quite naturally these anthropometric observations on *criminals* attracted public attention, and a whole literature appeared, particularly in Italy, on the subject of the anthropology of criminals. The most prominent writer of this school was Cesare Lombroso, who in his

[1] J. H. Baxter, *Statistics, Medical and Anthropological . . . of over a Million Recruits,* Washington, 1875.

[2] B. A. Gould, *Investigations in the Military and Anthropological Statistics of American Soldiers,* New York, 1869.

[3] Perozzo, "Sulle curve della statura degli iscritti misurati in Italia," *Annali di Statistica,* ser. II, Vol. 2, 1878. See further, the introduction to Ridolfo Livi, *Antropometria militare,* Parte I. Roma, 1898.

[4] Alph. Bertillon, "Une application de l'anthropométrie," *Annales de démographie internationale,* V, 1881.

chief work, *L'uomo delinquente*,[1] gave a vast compilation of observations, made by himself or others. Unfortunately he was not of sufficient statistical accuracy, his conclusions frequently being anything but sound. His theory on the origin of crime will be mentioned below.

Problems of *heredity* had only occasionally been touched upon before the middle of the century; for instance, *tuberculosis*, as had been done by Louis (see above, art. 53), and *cretinism*.[2] Sound results were hardly obtainable in a period when the conception of the influence of infection and of inherited disposition was still very vague; moreover, medical statistics on the whole were still in a primitive condition. But when in 1859 Ch. Darwin had published his *Origin of Species*, statistical investigations on heredity naturally followed. Darwin's theory made a deep impression on his cousin, Francis Galton (1822–1911), who at once became a convert to his views. In 1869 Galton published his first work on problems of heredity,[3] in which he claims to be the "first to treat the subject in a statistical manner." In his discussions he refers to the "law of deviation from an average," referring to Quetelet and quoting results of measurements of chest and height of men. The investigations in this first attempt are still lacking in clearness, but they formed the starting-point for further contributions from himself, and for numerous experiments made at his initiative, all of which resulted in the appearance in 1889 of his *Natural Inheritance*.[4] In this work are to be found the elements of the "law of regression," which afterwards played so conspicuous a part in the theory of statistics.

[1] C. Lombroso, *L'uomo delinquente in rapporto all'antropologia, alla giurisprudenza ed alle discipline carcerarie*, 4th ed., Torino, 1889; 1st ed., 1871–6.

[2] Brierre de Boismont, "Analyse du rapport de la commission créé par S. M. le roi de Sardaigne, pour étudier le crétinisme," *Annales d'Hygiène publique*, 43, 1850.

[3] Francis Galton, *Hereditary Genius*, London, 1869; see also his *Autobiography: Memories of my Life*, 1908, and Karl Pearson, *The Life, Letters and Labours of Francis Galton*, Vols. I–III, Cambridge, 1914–30.

[4] Francis Galton, *Natural Inheritance*, London, 1889.

The epoch here concerned can therefore also in this respect claim to be a point of preparation for the rich development of the following decades.

The fact that Galton in his first work on heredity emphasised the law of deviations is a sign of the progress in the evolution of statistics. By this time actuaries and statisticians began to become aware of the advantage to be derived from the calculus of probability. It was felt important, for instance, to be able to calculate the risk of the business of a life-office, and in all statistical investigations the question was how to judge of the accuracy of the results. It may be objected that in his work Galton did not reach the full conclusions of the theory, but one step more would have sufficed in order to master the tests of the calculus of probability as derived by Poisson and as recommended, although mostly in vain, by Gavarret (see above, art. 54).

74. Regarding the risk of the life-offices a remarkable treatise by Bremiker appeared in 1859.[1] He used the method of least squares to find the mean error of a life-assurance or an annuity, as well as the combined risk for all the policies of a life-office, under the supposition that the rates of mortality did not change, and that also the rate of interest was constant. The problem was later treated by the Danish mathematician J. P. Gram, who used the continuous method in order to derive his results.[2] The present author dealt with the risk by fire insurance, among other questions discussing the effect of a fusion of two societies.[3]

The problem of the *adjustment* of a mortality-table by means of the method of least squares may be mentioned here. It was treated by two Danish mathematicians, L. Oppermann and T. N. Thiele. The former adjusted the experience of the Government life insurance according to a formula, which, however, he did not

[1] C. Bremiker, *Das Risiko bei Lebensversicherungen*, Berlin, 1859.

[2] J. P. Gram, "Om Middelfejl paa Værdien af Livsforsikringer," *Tidsskrift for Mathematik*, 5 R 6, København, 1889.

[3] Westergaard, "Das Risiko bei Feuerversicherungen," *Assecuranz-Jahrbuch* V, Wien, 1884.

publish himself. His method was afterwards explained by J. P. Gram.[1] Thiele used the same material as Oppermann as basis of another formula for the mortality-table.[2] It may be added that several mechanical adjustments were also tried during this period, with more or less success.

The tests of the calculus of probability for the *validity* of *statistical conclusions* were treated by various authors in Germany and England. In 1867 Wittstein published a treatise on mathematical statistics, a denomination which, after his day, came into general use.[3] Supposing that the probability of an event is known, he calculates the probable error for a given number of cases, for instance, survivors after a certain lapse of time. In a lecture in 1868 W. Lazarus entered upon the question,[4] chiefly confining himself to the "skewness" of the binomial law, for which he derives a formula, whereas Wittstein had left this side of the problem alone. But a deeper investigation was made by Woolhouse (1872).[5] This article was less remarkable for its general introduction on statistics, for which he was even reproached for having too freely copied passages of Quetelet's Theory of Probabilities, than for the concluding part, in which he gives the solution of various problems, without, however, entering much into the proof of the theorems which he applies. He shows how to solve the two main problems, viz. *firstly*, to find the probability that the events concerned will occur within certain limits, if the probability is *known*, and *secondly*,

[1] J. P. Gram, "Om Udjevning af Dødelighedsiagttagelser og Oppermanns Dødelighedsformel," *Tidsskrift for Mathematik*, 5 R 2, 1884.

[2] T. N. Thiele, *En mathematisk Formel for Dødeligheden*, Kjøbenhavn, 1871.

[3] Wittstein, *Mathematische Statistik und deren Anwendung auf National-Oekonomie und Versicherungs-Wissenschaft*, Hannover, 1867. Translated by T. B. Sprague in *Journal of the Institute of Actuaries*, XVII, 1872, under the title, "On Mathematical Statistics and its application to Political Economy and Insurance."

[4] Wilhelm Lazarus, "On some problems in the Theory of Probabilities." Translated by T. B. Sprague and J. Hill Williams in *Journ. Inst. Act.*, XV, 1870.

[5] W. S. B. Woolhouse, "On the Philosophy of Statistics" (*Journ. Inst. Act.*, XVII, 1872).

to find the corresponding probability, if the frequency of the events be observed first, and the number of the new set of events has to be calculated from this material. It was naturally life insurance which he had in view, but he remarks that the methods of computation which he has shown will apply generally to all kinds of statistical inquiries. But like so many other authors who have written on problems of the theory of probability, he is not much interested in the question whether the statistical phenomena observed will obey the binomial law or not; his position in this respect is chiefly dogmatic.

Here we meet with a deeper insight in the remarkable contributions by W. Lexis,[1] of which one was quoted above. In the first of these treatises Lexis, after having dealt with the same problems of the movements of a population which Zeuner and Knapp had treated, proceeds to the problems of the theory of probability, including Poisson's formula concerning the difference between two values of frequency, and he shows how these calculations can be done practically. The second treatise contains various investigations as to whether there is harmony between the statistical phenomena and the law of error or not. In fact, he confesses that he has only found one field where the theorems of the calculus of probability are in harmony with the observations, viz. the sex-proportion of new-born children. Mostly he finds a more or less significant disharmony, the dispersion from the mean being as a rule greater than would be expected on the basis of the theory. As a test he uses a double calculation of the probable error, viz. firstly on the basis of the observed frequencies of the events, and secondly by the method of least squares. If the dispersion is greater than expected, the latter quantity will exceed the former. He

---

[1] W. Lexis, (1) *Einleitung in die Theorie der Bevölkerungsstatistik*, Strassburg, 1875; (2) *Theorie der Massenerscheinungen der menschlichen Gesellschaft*, Freiburg, i, B, 1877; (3) "Ueber die Theorie der Stabilität statistischer Reihen," *Jahrb. für Nationalökonomie und Statistik*, 32, 1879.

reverts to this theory of dispersion in the above-quoted article published two years later.

The proposal to make positive investigations as to the practical applicability of the law of error to various fields was of great importance. The road was opened for many-sided work in this respect.

The result of these efforts was more favourable than according to Lexis, since it proved that a systematic preparation of the material in many branches of human statistics would lead rather easily to an approximation to the law of error.[1]

Better known than Lexis' theory of dispersion are perhaps his investigations on the *normal age* at death. He tries to show that the deaths according to a mortality-table in advanced years will centre according to the law of error around a certain year, whereas in the earlier part of life a number of cases of death occur, from various reasons, without any connection with the law of error. He found, for instance, as the normal age for males, in *Norway* 74 years, in *England* 72, and in *Belgium* 67 years.

On the threshold of the following epoch a worthy supplement came from F. Y. Edgeworth in a lecture which he delivered at the Jubilee meeting of the Statistical Society in June, 1885, where in a lucid way he discussed the main problems in the theory of statistics, showing in each case how to find the limits of the deviations from the mean.[2]

In this way he treats, for instance, a series of anthropometric observations, asking whether a difference of height found between two groups of men might be considered accidental or not. Birth statistics formed another subject, chiefly with regard to sex-proportion: several investigations on the Hofacker-Sadler hypothesis,

---

[1] Westergaard, "Zur Theorie der Statistik," *Jahrb. für Nationalökonomie und Statistik*, Neue Folge, X, 1884.

[2] F. Y. Edgeworth, *Methods of Statistics*, Jubilee Volume of the Statistical Society, London, 1885. In the first edition of the present author's work, *Die Lehre von der Mortalität* (1882), the mean error to a large extent is used as test.

published at the time, for instance, by Goehlert,[1] would have been laid aside at once if the test advocated by Edgeworth had been in use. The lecture also contains some curious examples of conformity with the law of error, such as the number of wasps entering or issuing from their nest, and the attendance at a club of its members.

In the very short subsequent discussion the author was asked rather naïvely whether in his paper he had brought forward any new methods. Edgeworth replied modestly that he did not know that he had offered any new remarks, but perhaps they might be new to some readers. This was true, but at all events his paper contained a comprehensive review of the results which the theory of statistics had hitherto attained, and at the same time it gave, so to speak, a programme for further exploration of the law of error and for the regular application of the tests now within the range of the statisticians but which generally had only been reluctantly accepted. It was evident now that the theory of statistics had a really scientific basis, even though dilettantism was not yet extinct.

75. If rational principles could gain ground but slowly in the strictly so-called *vital statistics*, this was still more the case with *moral statistics*. This branch shared the fate of *medical statistics* in being rather ill-defined. At all events there was general assent that this branch of statistics embraced certain subjects, such as suicide, crime, prostitution, illegitimate births and divorces, and in this epoch a vast literature on such subjects saw light. Small wonder that the results with regard to the regularity of many phenomena had been received with general attention. But the popularity of the subject attracted many authors who, while able to speak with authority on other subjects, as statisticians

---

[1] Goehlert, *Untersuchungen über das Sexualverhältniss der Geborenen* (1854), *Statistische Untersuchungen über die Ehen* (1869), *Die Geschlechtsverschiedenheit der Kinder in den Ehen* (1881); see I. Wedervang, *Om Seksualproporsjonen ved Fødslen*, Oslo, 1924, pp. 23 *sq.*

were only dilettanti. Their methods were often obsolete, and their material often collected without critical
sense. Further, the interpretation of the results was
in many cases ill-founded. In fact most of the observations could only be looked upon as more or less *indirect*
symptoms. It may, for instance, be possible to find the
number of persons belonging to a certain church, but
this cannot, of course, give any evidence as to their
inner religious life.

Still, it was not impossible to arrive at safe conclusions
on many points of interest. *Suicides* were in this period
frequently the subject of investigation: here the same
methods can be applied as to all other causes of death
(relative frequency of suicides according to age, sex,
season, etc.), and important results may be obtained in
this way, even though it will be difficult or even impossible to consider the observations in other directions,
as, for instance, the *motives* of the suicides. And even
though *illegitimacy* is only a symptom of the morality of
a population, it is possible to obtain valuable results
concerning the numbers born out of wedlock in various
classes of society, etc. The efforts of Neison, Guerry
and others to study criminality in connection with
education and other circumstances were not very successful, but the phenomena in this field permitted many
sober calculations—corresponding, for instance, to the
construction of a life-table—which might prove useful,
by discussions as to the frequency of crime, particularly
if individual observations on relapses are obtainable.

The great difficulty was to get an unbiassed interpretation of the facts. Too often the authors were
misled by their individual view; they would often—
unconsciously—find their own personal opinion supported by facts which their opponents would interpret
from quite another standpoint.

This will be seen by a survey of the statistical literature of the period concerned. Much material for a
study of this kind will be found in various statistical
periodicals, as in the *Journal of the Statistical Society of*

London, and not least in the Italian *Annali di Statistica*, abounding in original papers on subjects of moral statistics as well as abstracts and reviews.[1]

And just as Süssmilch in the eighteenth century published a compendium of the statistical knowledge of his time, so Alexander von Oettingen (Professor of Divinity in Dorpat) a century later issued a corresponding work on the more limited subject of moral statistics.[2]

He shared with many of his contemporaries the lack of deeper statistical insight,[3] but in spite of this defect his work is an honest attempt to guide the reader in all the various chapters of moral statistics.

In the period concerned the problem of *free will* quite naturally came under vivid discussion. The theories expounded by Quetelet and others (see above, art. 58) had made a deep impression, which applied in an even higher degree to Ad. Wagner's work on the regularity in human actions (1864).[4] The historian H. T. Buckle was a declared supporter of Quetelet's view (in his history of civilisation in England).[5] As a matter of fact, opinions were much divided. Many statisticians explained the facts in the same spirit as Quetelet, others took the opposite standpoint. To the former belonged the Italian criminological school, founded by C. Lombroso, to the latter group Oettingen, Drobisch and Knapp.[6]

---

[1] For instance, Enrico Ferri, "Studi sulla criminalità in Francia," *Annali di statistica*, Ser. 2, Vol. 21, 1881. Morselli, "Il suicidio Saggio di statistica morale comparata," ibid., Vol. II, 1880 (the work itself appeared the year before).

[2] Alexander von Oettingen, "Die Moralstatistik und die christliche Sittenlehre," *Versuch einer Socialethik auf empirischer Grundlage*, Erlangen, 1868–73, 3rd ed., much enlarged, appeared under the title, "Die Moralstatistik in ihrer Bedeutung für eine Socialethik," ibid., 1882.

[3] This, for instance, prevented him from a proper valuation of Goehlert's investigations on the sex-proportion (l.c., 3rd ed., p. 77); see also his remarks on the frequency of suicides according to marital condition (p. 776).

[4] Ad. Wagner, *Die Gesetzmässigkeit in den scheinbar willkürlichen menschlichen Handlungen*, Hamburg, 1864.

[5] Buckle, *History of Civilisation in England*, 1857.

[6] M. W. Drobisch, *Die moralische Statistik und die menschliche Willensfreiheit*, Leipzig, 1867. G. F. Knapp, *Die neuern Ansichten über die Moralstatistik*, Jena, 1871.

Lombroso found it proved by statistical and anthropological investigations that crime is a natural phenomenon which to a great extent depends on the organism and education or surrounding circumstances. Enrico Ferri speaks in the same tone, maintaining that the frequency of crimes depends necessarily on natural and social circumstances, combined with hereditary and acquired individual propensities, and he looks upon these results as influencing the practical evolution of criminology.[1]

As a contrast to this naturalistic view, Oettingen maintains that a human being does not only obey natural influences, but that at the same time he has a personal life, in free responsibility, so that normally he will be able to arrive at an ethical conviction which permits him to form resolutions in full freedom.[2] Drobisch is less sharp in his opposition to Quetelet and his followers; he arrives at the result that even though there is no spontaneous, absolute free will, human actions are regulated by motives, which not only depend on outer, surrounding circumstances, but on inner causes: a human being having the choice between following his instincts and being guided by his personal character.

The discussion never reached a final conclusion, and even now—two generations later—there will be widely different opinions as to the interpretation of the facts in moral statistics. But statisticians will nowadays generally agree with Knapp in his conclusion that the whole problem of the human will does not belong to statistics but to philosophy, statisticians having it in their power to render philosophy practical services but not to solve the problem itself. This, in fact, will hold good with many other sciences where statistics can give valuable help in collecting observations and controlling the results. The enthusiasm which a century ago was willing sometimes even to crown statistics as the queen

---

[1] Confer with regard to the Italian school, A. Gabaglio, *Teoria generale della statistica*, Parte storica, 2nd ed., Milano, 1888, pp. 284 *sq.*

[2] Oettingen, l.c., 3rd ed., p. 805.

of all sciences, no longer exists; it has been replaced by a more sober view. But on the other hand the usefulness of statistical investigations based on rational principles can now be more fully appreciated, and in the following two generations practical and theoretical statistics were able to win a more consolidated position than was dreamt of in the era of enthusiasm.

CHAPTER XVII

THE LAST DECADES OF THE
NINETEENTH CENTURY

76. THE evolution sketched in the preceding chapters evidently gave statistics a much more solid basis than before. And in the last two decades of the past century further progress can be recorded—in official statistics as otherwise, particularly also in theory.

It is outside the plan of the present treatise to give a detailed description of the progress of official statistics all over the world. For a detailed study of this kind, however, the material is easily accessible, partly in the literature which has been quoted above, partly in numerous monographs pertaining to the condition of statistics in the various countries, in official publications, in statistical journals, etc. Some general remarks may suffice here, illustrated by a few facts where it is found useful.

A student of publications of statistical institutions at the close of the past century will early obtain the impression that a true scientific spirit is generally prevailing. The emulation between the various countries and the mutual teaching of the statisticians undoubtedly had not been in vain. The statisticians had learnt what questions it was possible to put, and how to formulate them. The technical process in dealing with the material was naturally improving, and the statistical service better organised.

The general evolution of political and constitutional life was favourable for the progress of official statistics. Everywhere statistical information was required. Self-governing institutions and other authorities often introduced a statistical service of their own, and new central statistical organisations were established, the result on the one hand being a *division of labour*, on the other, an effective *centralisation*.

In the *division of labour, municipal statistics* took a prominent part. The administration of municipalities quite naturally required collection and publication of various statistical data. This was frequently done by different departments within the administration till at last the desirability for centralisation was felt. In *Paris*, as we have seen, statistical publications were issued rather early in the *century*, six volumes of *Recherches statistiques* appearing 1821 to 1860. Here in 1879 a central office for municipal statistics was created (Bureau de Statistique de la Ville de Paris). The first director was L. A. Bertillon, well known as author of statistical investigations, particularly on medical statistics.[1] His son, Jaques Bertillon, succeeded him after his death in 1883. *Vienna* had already a municipal statistical bureau in 1862, *Budapest* in 1869. The latter, till 1906, was directed by the energetic Hungarian statistician, J. Kőrösi; *Amsterdam* got a municipal bureau in 1894 (Falkenburg); in *Scandinavia* a local bureau was established 1883 in *Copenhagen*, 1887 in *Christiania* (*Oslo*), and *Stockholm* followed in 1905.

In the period concerned municipal statistics also made great progress in *Germany*. At the beginning of 1880 there were in the Empire twelve municipal statistical bureaux.[2] During the following two decades the number was doubled, and in 1900 four more bureaux of this kind were added. The best known among them is the *Berlin* office, established in 1862, which—particularly under the direction of R. Böckh (1874–1902)— won a great reputation, largely on account of the rich contents of its year-books. Among several others the

---

[1] *La vie et les œuvres du Dr. L. A. Bertillon,* Paris, 1883 (published by his sons).

[2] M. Neefe, "Die städtischen statistischen Aemter," *Handwörterbuch der Staatswissenschaften,* 3rd ed., VII, 1911, pp. 893 *sq.*

Leipzig bureau also deserves mention; it was founded 1867 and directed by G. F. Knapp till 1874.

These municipal institutions were charged with the centralisation of statistical information on various subjects such as local finance, police, etc. They rendered valuable service particularly with regard to vital statistics and hygiene. Knapp and Böckh's contributions in this field were of considerable value in a theoretic respect.

Perhaps even more important in regard to *division of labour* is the progress of *labour statistics* (see above, art. 63). The growing interest in social problems would naturally lead to energetic efforts to obtain more profound statistical information concerning the condition of the working classes. The first attempt to create an institution charged with this subject was made in *Massachusetts*, 1869.[1] The duty of this bureau of statistics of labour was to "collect, assort, systematize, and present statistical details relating to all departments of labor." It took up several subjects of importance in this connection; for instance, wages, hours of labour, and working conditions of the wage-earners, statistics of manufactures, etc. Other States within the Union followed, *Pennsylvania* in 1872, *Ohio* in 1877, and in the last twenty years of the century a large number of similar institutions were created. The *Federal Government* also established an institution for labour statistics. After some desultory efforts to secure data relating to wages a Bureau of Labour was created in 1884, and four years later it was made an independent department.[2] The Act of 1888 charged the department with several important duties; for instance, investigations on all controversies and disputes between employers and employees. It also had to ascertain what products were controlled by trusts, etc., to report as to the effect of the custom laws and to make investigations as to "the general

---

[1] Ch. F. Gettemy, "The Work of the Several States of the United States in the Field of Statistics," *The History of Statistics*, 1918, pp. 692 *sq.*

[2] John Cummings, "Statistical Work of the Federal Government of the United States," l.c., pp. 633 *sq.*

condition as far as production is concerned, of the leading industries in the country."[1] These and several other subjects were taken up in course of time. The Department was conducted for many years by Carroll D. Wright.

About the same time as the United States, *Great Britain* founded a Labour Department (1886) which developed into a large and important centre of information as to wages, employment and conditions of the wage-earning classes generally.[2] A journal, *Labour Gazette*, was published from 1893.

Several other countries took part in the energetic competition to secure information in this field. In *Sweden*, for instance, labour statistics were given a place in the State administration in the year 1897, a separate branch for this object being organised within the Board of Trade in 1903.[3] In *Austria* an office for labour statistics was installed at the Board of Trade in 1898.[4] In *Belgium* a similar institution was created in 1894.[5] In the *German Empire* a Commission for labour statistics was established in 1892.

77. In *Germany* we meet with a healthy competition between the Department for the Empire, which, as we have seen, was created in 1872, and the bureaux for the separate States. Two well-known names are associated with the *Bavarian* bureau, that of F. B. W. Hermann, who directed it 1830–68, and that of G. v. Mayr (1869–79); moreover, in the period treated here this institution played a prominent part in German statistics. The same holds good with regard to the *Prussian* bureau, which has frequently been mentioned in preceding

[1] E. R. L. Gould, "The Progress of Labour Statistics in the United States," *Bull. de l'Institut International de Statistique*, VI, 1, 1892.

[2] A. Baines, "The History and Development of Statistics in Great Britain and Ireland," *The History of Statistics*, 1918, p. 375.

[3] E. Arosenius, "The History and Organization of Swedish Official Statistics," l.c., p. 558.

[4] Robert Meyer, "The History and Development of Government Statistics in Austria," l.c., pp. 103 *sq.*

[5] Armand Julin, "The History and Development of Statistics in Belgium," l.c., p. 136.

chapters. E. Engel was its director, 1860–92. In *Saxony*, V. Böhmert (1875–95), and after him, A. Geissler (1895–1902), were the leaders of the State bureau.

It is not without interest to follow the evolution of official statistics in larger commonwealths. A few facts may be added to those mentioned above (art. 63). Soon after the Confederation in 1867 *Canada* began to make progress with regard to official statistics. The Dominion Parliament provided for a general census in 1871, and in a later Act (1879) for a permanent decennial census, and for collecting and publishing vital, agricultural, commercial, criminal and other statistics; a permanent Census and Statistics Office was organised in 1905.[1] As to vital statistics, however, progress was rather slow. Some of the provinces had passed modern registration laws, but for various reasons the data were very deficient, and although the unification of registration and of vital statistics was felt as a necessity, no real progress was made in this field till the present century.[2]

In *Australia* conditions seemed more favourable. Although the *Commonwealth of Australia* did not come into being till New Year's Day, 1901—a central organisation of Australian statistics being established by the Census and Statistics Act of 1905—official statistics made great progress in the decades preceding this event.[3] In the year 1873, as we have seen, the State of *Victoria* appointed a "Government Statist," who in addition to his statistical duties was also charged with the registration of births, deaths and marriages. In 1886 a statistical office was established in *New South Wales*—the first Australian office which was altogether separated from other offices. Since its establishment it has issued a long series of valuable statistical reports and year-books. Several conferences of the State statis-

---

[1] E. H. Godfrey, "History and Development of Statistics in Canada," l.c., pp. 180 *sq.*

[2] Robert R. Kuczynski, *Birth Registration and Birth Statistics in Canada*, Washington, 1930, pp. 4 *sq.*

[3] G. H. Knibbs, "The History and Development of the Statistical System of Australia," *The History of Statistics*, 1918, pp. 57 *sq.*

ticians took place, often resulting in real improvements, as was the case with the census of 1891.

In the *United States of America* there were great difficulties. The decentralisation ruling here was on the whole unfavourable to official statistics.[1]   For each decennial census a bureau was established, but there was no permanent federal bureau to keep account of population statistics. F. A. Walker, who was the leader of the census 1870 and 1880, made great efforts in vain to have an institution of this kind established. When the bureau for the twelfth census (1900) was created, the Acts carefully provided that "nothing herein contained shall be construed to establish a census bureau permanent beyond the Twelfth Census." Some years later, however, this was realised, an Act, 1902, making this bureau permanent. As we have seen, Federal statistics had great difficulties with regard to births and deaths, nor did the various States make any great progress in this respect.   In spite of several efforts there was a serious lack of system in this field,[2] although in other branches of statistics real progress was made. The Federal Bureau of Immigration, for instance, established in 1891, had immigration statistics to deal with, the Department of Agriculture, agricultural statistics, the Interstate Commerce Commission (1887) had among other subjects railway statistics.   And the above-mentioned institutions for labour statistics were a conspicuous step forward.

Still greater were the difficulties in the *Russian Empire*.[3] After many years of deliberation the first general census of the population was taken in 1897. On account of the low degree of education among the agricultural

---

[1] John Cummings, "Statistical Work of the Federal Government of the United States," l.c., pp. 677 *sq*. See also a report in the Publications of the. American Statistical Association, VII, 1901.

[2] Fr. L. Hoffman, *The Vital Statistics of the Census of 1900*, Publications of the American Statistical Association, VIII, 1902–3. Ch. F. Gettemy, "The Work of the Several States in the Field of Statistics," *The History of Statistics*, 1918, pp. 691 *sq*.

[3] A. Kaufmann, "The History and Development of the Official Russian Statistics," l.c., pp. 473 *sq*.

population it was necessary to prescribe oral interrogation in the villages, but there was a serious lack of properly qualified persons to act as enumerators. Nor was it easy to get reliable details in the towns. The process of compilation and the further work was remarkably slow: the general results were not published until 1905. A similar delay took place with regard to the statistics of births, deaths and marriages; the data for 1873, for instance, remained unpublished for nine years. The conditions of official statistics appear to have been more favourable in *British India* with its enormous population.[1] Here agricultural statistics date from 1881; in the same year a general census was taken; it was repeated at ten-years intervals.

78. As to the *Centralisation* of statistics much will, of course, depend on tradition. *England*, for instance, has the historic division of labour between the Board of Trade and the Registrar-General of Births, Deaths and Marriages. *France* was in a similar position; the Bureau de la Statistique Générale was chiefly in charge of population and vital statistics, the Ministry of Justice looked after judicial statistics, etc.[2] Yet since 1878 the general bureau represented to a certain extent a unity of the various statistical branches in so far as it was charged with the preparation and publication of an *Annuaire Statistique*.

In small countries such as *Denmark*, where official statistics were reorganised in 1895, or *Norway*, where the Central Statistics Office was organised in 1875, it was easier to centralise statistics by referring to the central office a large number of subjects. Similar attempts were made in *Holland* although not without difficulties.[3]

---

[1] A. Baines, "The History and Development of Statistics in British India," l.c., p. 421.

[2] F. Faure, "The Development and Progress of Statistics in France," l.c., pp. 288 *sq.*

[3] C. A. Verrijn Stuart, "The History and Development of Statistics in the Netherlands," l.c., pp. 435 *sq.*; see also by the same author, "La réorganisation de la statistique dans les Pays-Bas," *Bull. de l'Institut international de statistique*, XII, 1900. H. W. Methorst, "Zur Geschichte der niederländischen Statistik, *Allg. Statistisches Archiv*, 6, 1904.

Vain attempts were made in 1879 to create an official and permanent central statistical bureau. The above-mentioned Union for Statistics, however, received an allowance, and with the assistance of the Municipality of Amsterdam it established in 1884 its own statistical institute, in the university buildings of Amsterdam. Later on, this institute was dissolved and a Central Commission for Statistics was established (1892); in addition to this commission a Central Bureau of Statistics was finally created (1899), to which various subjects of statistics hitherto compiled by the different departments were transferred.

Whether centralisation was realised to the extent of a single institution having charge of the chief statistical subjects, or whether the opposite was the case, progress with regard to the *manipulation* of statistical observations naturally caused a centralisation within the particular branch, the compilation being gradually removed from local authorities to the central institution. Quite naturally the old-fashioned machinery gave way to a more mechanical process. We have a typical instance in the history of official statistics in *France*. Here, as mentioned above (art. 63), the individual report (bulletin) was introduced by the census of 1876, the evident result being more accuracy in enumeration and greater ease in compilation. But it was still a drawback that each of the 30,000 communes had to make the compilation, numerous mistakes being unavoidable, even though these reports were carefully revised by the *préfectures*. In 1881 a step forward was taken, a fixed date of the census being ordered: whereas previously the duration of the enumeration was indeterminate and variable, it was now decided that the census should be taken on a certain day, every person being included who had spent the previous night in the commune. But the statistical experts claimed further progress, viz. a system of central compilation. This at last was realised by the census in 1901, a special department being established for this purpose.

The usual way of compiling statistical data from the lists originally consisted in the laborious counting of homogeneous facts, marking, for instance, each individual case by a dot or a dash. When the material is large this process will often be inconvenient and errors can hardly be avoided. An important step forward will be to put down the details on individual cards which can subsequently be arranged in homogeneous groups. This method was first applied by insurance societies, especially for investigations on the mortality of insured lives;[1] later on it came into use in official statistics particularly in the case of census results and the movements of population. Here the last decades of the past century appear to have been fertile. By the census of 1880, for instance, the city of *Vienna* had individual cards filled in from the lists. In *Norway* they were used already in 1876. The same was the case with the *German* industrial census, 1882. In *Prussia* the individual card was introduced already in 1874 by the registration of births, deaths and marriages, and several *German* States soon followed the example. *Italy* first applied the system for causes of death, in certain cities (1881), later extending the system. In *Switzerland* the system came into use in 1876 with regard to the movements of population, and it was applied in 1887 to the statistics of transatlantic emigration; the census of 1888 was compiled by means of individual cards.[2]

This technical progress is naturally followed by a concentration of statistics, the greater part of the compilation being placed in the hands of the central institution instead of the communes. The process of compilation will again

---

[1] As, for instance, the mutual *English* investigation which was resolved in 1862 (*The Mortality Experience of Life Assurance Companies collected by the Institute of Actuaries*, London, 1869) and the *German* experience, *Deutsche Sterblichkeits-Tafeln aus den Erfahrungen von dreiundzwanzig Lebensversicherungs-Gesellschaften*, Berlin, 1883, pp. xxxiii *sq.*; this investigation was resolved in 1868.

[2] H. Rauchberg, "Uebersicht über den Stand und die neuesten Fortschritte der Technik auf dem Gebiete der Bevölkerungsstatistik," *Allg. Statistisches Archiv*, i, 1890–1. The same author, "Die Deutsche Berufs- und Betriebszählung vom 5 Juni, 1882," *Statistische Monatschrift*, Wien, XIV, 1888.

be facilitated if the various facts which have to be counted are indicated by means of a system of perforation of the cards, a hole in a certain place of the card announcing a certain fact, such as age, marital condition, sex, etc. And as the last step forward the *counting machine* is taken into use.[1]   The *Hollerith* electric machine was introduced for the census 1890 in the United States, and in the same year *Austria*, as the first European State, applied this invention.   It was used in *Norway* for the compilation of statistical observations on income and wealth for the year 1891.[2]   This invention has been of great practical help, the electric machinery adding considerably to the quick and exact compilation of the facts.   The application of this invention during the present century in a large number of countries has been one of the most important features of modern practical statistics.

79. We have seen in a preceding chapter how the *International Statistical Congress* was established, and how this institution flourished till it degenerated and at last came to an end (see art. 62).   The need of a co-operation of this kind was felt much less fifty years ago than in the middle of the past century.   Statisticians had easy access to scientific and social intercourse whether there were congresses or not, and there was no hindrance to an active and lively co-operation between the statistical offices all over the world.   Yet it was often felt desirable among statisticians to create an institution which in a more effective way than the Congress might unite representatives of practical and theoretic statistics.   This was realised in the year 1885 by the foundation of the *International Statistical Institute*.   In this year the Statistical Society of London (founded in 1834) celebrated its

---

[1] H. Rauchberg, "Die elektrische Zählmaschine und ihre Anwendung bei der österreichischen Volkszählung," *Allg. Statistisches Archiv*, 2, 1891.   Confer further a report in the *Bulletin de l'Institut international de statistique*, VI, 1, 1892, and Ernst Blaschke, *Vorlesungen über mathematische Statistik*, Leipzig und Berlin, 1906, pp. 257 *sq.*

[2] A. N. Kiær, "The History and Development of Statistics in Norway," *The History of Statistics*, 1918, p. 459. See "Indtægts og Formueforhold i Norge," *Norges officielle Statistik*, III, No. 255, Kristiania, 1897.

jubilee. One of the objects of this meeting was to "consider the possibility of establishing an International Statistical Association." The *Austrian* statistician, Neumann-Spallart, gave an outline of the proposed association. The rules were accepted with some amendments, and the society (l'Institut International de Statistique) was constituted with Sir Rawson W. Rawson as president and L. Bodio as general secretary. One obstacle, however, had to be removed. The German statisticians, Becker and Blenck, the chiefs of the official statistics of the German Empire and of Prussia, hesitated to accept membership in the fear that their duties as members of the Institute might interfere with their official position. This was not without connection with the events which led to the death of the Congress. After some correspondence, however, they were assured that the new Institute was a thoroughly free Association which would in no way try to bind the respective Governments by their resolutions. Actually the Institute has been true to this principle ever since its foundation.

The first meeting of the Institute should have taken place in *Rome* in the following year, but it had to be postponed to 1887 on account of the cholera epidemic which had attacked Italy. At this meeting the rules were amended and somewhat simplified. According to the rules of 1887 the maximum number of ordinary members was 150. This limitation gave the sessions of the Institute a much greater working power than the immense assemblies which the later congresses could boast of. The list of members was very promising for a good future for the Institute, several of the best known statisticians and economists being on the roll. Among the members enlisted in 1887 we find several men prominent in active statistical service. Among theorists we meet, for instance, Galton, Knapp and Lexis. That E. Engel, G. v. Mayr and A. Soetbeer were among the statisticians who joined was a matter of course. A fairly large number of famous older or younger economists was enlisted; for instance, among Germans, Brentano,

Roscher, Schmoller and Wagner; from other countries, Goschen, Laveleye, Léon Say and Walras.

The Institute published a Bulletin, the first volume appearing in 1886–7.[1] This at once became a treasure for students of practical and theoretical statistics. The reports of the sessions are printed in the bulletins, as well as a number of monographs, obituaries, bibliographical lists, etc. Naturally international statistics claimed a large space. The well-known French statistician Levasseur contributed to the first volume an elaborate compilation concerning area and population of the countries all over the world. A second edition in co-operation with Bodio appeared later, bringing the numbers up to the year 1900.[2] The initiative of L. Bodio manifested itself in international statistics of the movements of population.[3] Problems of criminal statistics were discussed, for instance, with regard to individual cards which in this period began to be applied, here with particular application to recidivism.[4] The *Norwegian* statistician A. N. Kiær contributed to international shipping statistics, Craigie to agricultural statistics, etc. The condition of the working classes claimed a particular interest. Disciples of Le Play entered on his system of monographs concerning individual families.[5] *Belgian* statistics on the budgets of working men attracted great interest;[6] in 1895 E. Engel published an elaborate treatise on this subject.[7]

---

[1] *Bulletin de l'Institut international de statistique*, Rome, I, 1886–7.

[2] E. Levasseur, "Statistique de la superficie et de la population des contrées de la Terre," *Bull.* I–II, 1887. E. Levasseur et L. Bodio, *Statistique*, etc., XII, 1902.

[3] "Movimento della popolazione in alcuni Stati d'Europa e d'America," *Bull.* VII, 1894; X, 1897.

[4] *Bull.* XII, 1, 1900, p. 111.

[5] Urbain Guérin, "De la méthode de monographies de famille," *Bull.* III, 1888. E. Cheysson et A. Toqué, *Les Budgets comparés des cent monographies de familles*, V, 1890.

[6] See, for instance, Hector Denis, *Note sur les indices de la prospérité et spécialement sur les budgets ouvriers*, IV, 1890.

[7] Ernst Engel, "Die Lebenskosten belgischer Arbeiter-Familien früher und jetzt—Ermittelt aus Familien-Haushaltrechnungen und vergleichend zusammengestelt," *Bull.* IX, 1895.

Various theoretic problems came under discussion in the Institute: A. N. Kiær treated the problem of sampling,[1] Kőrösi the standard calculations in mortality statistics.[2] There are further investigations on fertility,[3] on divorces,[4] and on the nomenclature of causes of death.[5] R. Böckh published his classic treatise on Halley's table of mortality.[6] These are only instances of the many-sided contents of the Bulletin and of the discussions in the Institute.

80. The foundation of the International Statistical Institute was not the only sign of a co-operative spirit. As we have seen, English life-offices co-operated at an early date with regard to mortality experience. After the report of 1843 from 17 offices a second co-operative report followed, embracing the experience of 20 societies (1869). A third investigation which had great influence in the past generation was based on observations from 1863 till 1893.[7] G. F. Hardy undertook the work of graduation, Th. G. Ackland gave an account of the principles and methods. Altogether 66 companies co-operated. The above-quoted experience of 23 *German* life-offices, the compilation of which was resolved in 1868 and was completed in 1883, also enjoyed a high reputation. The graduation was undertaken by Zillmer.[8] In *America* 30 societies combined their

---

[1] A. N. Kiær, "Observations et expériences concernant des dénombrements représentatifs," *Bull.* IX, 1896.

[2] J. Kőrösi, "Mortalitäts-Coefficient und Mortalitäts-Index," *Bull.* VI, 1892. "Ueber die Berechnung eines internationalen Sterblichkeitsmasses (Mortalitäts-Index)," *Bull.* VIII, 1895.

[3] For instance, Raseri, "Les naissances en rapport avec l'âge des parents," *Bull.* X, 1897; R. Böckh, "Die statistische Messung der ehelichen Fruchtbarkeit," *Bull.* V, 1890.

[4] R. Böckh, "Statistik der Ehescheidungen in der Stadt Berlin in den Jahren 1885 bis 1894," *Bull.* IX, 1899.

[5] Jaques Bertillon, "Sur une nomenclature uniforme des causes de décès," *Bull.* XII, 1900.

[6] R. Böckh, "Halley als Statistiker," *Bull.* VII, 1893.

[7] *British Offices' Life-Tables*, 1893. An Account of the Principles and Methods adopted in the compilation of the Data, . . . and the Construction of Deduced Tables. Assured Lives and Life Annuities. London, 1903.

[8] *Deutsche Sterblichkeits-Tafeln aus den Erfahrungen von dreiundzwanzig Lebensversicherungs-Gesellschaften*, Berlin, 1883.

experience.[1]   In *France* a co-operative investigation was finished in 1895.[2]   And in 1893, 19 *Scandinavian* societies took up the task.[3]

These investigations added much to the knowledge of the mortality of insured lives, supplemented as they were by numerous monographs from individual offices.   As a general result the chief actuarial problems could now be considered as solved, as, for instance, the correlation between the duration of the policy and the health of the insured.

As to the problem of the mortality among persons who on account of ill-health had been accepted only with increased premiums, the report of the twenty *English* Societies had already furnished instructive information, and the literature in the last decades of the nineteenth century contains numerous contributions in this respect. Another problem which now began to attract attention was that of finding the rates of mortality among persons whose application for assurance had been refused.   An attempt of this nature was made by a *Danish* life-office, an experiment of this kind having better prospects in a small country than in a large one.[4]   E. Blaschke tried to solve the problem by indirect methods, particularly through a study of the causes of death.[5]

Official mortality statistics also made conspicuous progress in this period.   The contributions of Zeuner, Becker and others, mentioned above (art. 72), had not been in vain.   Under the auspices of Becker a mortality-table for the *German* Empire was calculated.[6]   The

[1] Levi W. Meech, *System and Tables of Life Insurance*. . . .   Revised edition for 1886, New York.

[2] *Tables de mortalité du comité des compagnies d'assurances à primes fixes sur la vie*, Paris, 1895.

[3] *Dödlighetstabeller enligt nitton skandinaviska och finska lifförsäkringsanstalters erfarenheter.*   I Normala lif. Stockholm, 1906.

[4] Westergaard, "Strøbemærkninger om Livsforsikringsvæsenets Fremtidsopgaver," *Det gjensidige Forsikringsselskab "Danmarks" Livsforsikringsafdeling,* 1872–97.   Copenhagen, 1897.

[5] *Denkschrift zur Lösung des Problems der Versicherung minderwertiger Leben.*   Vienna, 1895.

[6] "Deutsche Sterbetafel, gegründet auf die Sterblichkeit der Reichsbevölkerung in den 10 Jahren 1871/72 bis 1880/81."   *Monatshefte zur Statistik des Deutschen Reiches*, November, 1887.

fluctuations of the population caused by migration were taken into consideration by means of the transatlantic emigration. Almost the same method was used in *Norway*.[1] R. Böckh made a corresponding calculation for *Berlin*, based on observations of removals to or from the city. He arrived at interesting results with regard to the mortality of illegitimate children, showing that the alleged better health of illegitimate than of legitimate children after a certain age, on account of selection, did not exist.[2] A. J. van Pesch calculated life-tables for Holland for the two decades 1870–80 and 1880–90, following each generation from year to year between two censuses by means of the death-lists, under the supposition that the surplus emigration could be distributed uniformly over the decade.[3] Less satisfactory were J. M. J. Leclerc's tables for Belgium.[4]

To these tables can be added numerous investigations on mortality in certain classes of society or in certain districts. Frequently the material did not permit of the improved methods which had been introduced in official statistics for a whole country. But the fact that (as mentioned above) various summary methods led to relatively reliable results, justified many of these investigations, so much the more as the results often show an inner harmony. Here the mortality in various occupations may be mentioned.

The work which William Farr had begun (art. 70) was in this period taken up by his successor, William Ogle,[5] and after him by John Tatham.[6] The former based his statistics on the death-lists, 1880–2, and the census, 1881, the latter on the corresponding material, 1890–2, both carefully considering the various sources

[1] *Livs- og Dødstabeller for det norske Folk efter Erfaringer for Tiaaret, 1871/72–1880/81.* Kristiania, 1888.

[2] See particularly *Statistisches Jahrbuch der Stadt Berlin,* 1885–6.

[3] *Sterftetafels voor Nederland,* Haarlem, 1885, and 's Gravenhage, 1897.

[4] *Tables de mortalité ou de survie et tables de population pour la Belgique,* Bruxelles, 1893.

[5] *Supplement to the Forty-Fifth Annual Report of the Registrar-General of Births, Deaths and Marriages in England,* London, 1885.

[6] *Supplement to the Fifty-Fifth Annual Report. . . .* Part II, 1897.

of error. Ogle applied a less specified age distribution than Farr (five classes of age only against eight), but on the whole the results seem to give a reliable picture of the differei.ces in health according to occupation. His above-mentioned attempt at finding the distribution after causes of death, by means of sampling, was a step forward, and he planned more complete statistics for the next decennial report. Tatham, who succeeded him in 1893, extended the plan, dividing the causes of death into twenty-four groups, and the ages into seven. As a means of comparison, both used a standard calculation with the age distribution in the general population as starting-point. The following decennial reports from this century have added much to the knowledge which was acquired by these investigations.

Several other contributions to the solution of the problem of the influence of occupation on health can be registered. Among these was an attempt made by J. Bertillon [1] with regard to the population in *Paris*. The census material for the *Danish urban* population served as basis for Th. Sørensen in investigations on mortality in various classes of society.[2] The mortality in the *medical* profession was treated by Geissler,[3] Johannes Karup und Gollmer.[4] Several interesting observations were made on the mortality of retired officers, for instance, with regard to the influence of selection during the first years of their being pensioned.

One of the problems which came under discussion in this epoch was that of the influence of *intoxicating liquors* on health. Indirectly the question was attacked by investigations on mortality among persons engaged in the *liquor* trade. The above-quoted decennial supple-

---

[1] J. Bertillon, *De la morbidité et de la mortalité par profession. Seventh International Congress of Hygiene*, London, 1891.

[2] Th. Sørensen, *Børnedødeligheden i forskjellige Samfundslag i Danmark*, Copenhagen, 1883. "De økonomiske Forholds og Beskjæftigelsens Indflydelse paa Dødeligheden I–II," ibid., 1884–5.

[3] Geissler, *Die Sterblichkeit und Lebensdauer der sächsischen Aerzte*, Leipzig, 1887.

[4] J. Karup und Gollmer, "Die Mortalitäts verhältnisse des ärztlichen Standes," *Jahrb. für Nationalök und Stat.*, N.F., XIII, 1886.

mentary reports gave valuable contributions in this respect and various life-offices treated their experience.[1] Several life-offices had sections for *abstainers* and were thereby enabled to compare the health of abstainers and non-abstainers. As a prominent example the United Kingdom Temperance and General Provident Institution of London can be quoted. The results, however, frequently met with criticism until more profound investigations during the present century could disperse the doubts.

The experience of *friendly societies* with regard to *sickness* was made accessible in this period in various reports. This was the case with two of the large *English* affiliated orders, and an elaborate report by William Sutton was published by the Chief Registry of Friendly Societies.[2]

Less fertile was the period with regard to the problem of *invalidity*. In various countries social legislation was setting in, but the material was still too new to give sufficient evidence. Yet some contributions on special groups of workers saw light. The best observations appeared in Germany, where Zimmermann and Zillmer continued the work of G. Behm concerning the invalidity and sickness among railway servants,[3] whereas Küttner published a treatise on the invalidity among coalminers.[4] Accidents in mines and on railways in England were treated by Neison, jun.[5]

81. Great difficulty frequently existed in obtaining reliable material for investigations on health. The prob-

---

[1] Gordon Douglas, Statistics as to the Mortality Experience among Assured Lives engaged in the Liquor Trade. Th. Wallace, "On the Rate of Mortality among Liquor Sellers" (*Transactions of the Act. Soc. of Edinburgh*, 1891).

[2] Copy of Special Report on Sickness and Mortality experienced in registered Friendly Societies, London, 1896 (*Sickness and Mortality Experience deduced from the Quinquennial Returns . . . for the years 1856 to 1880 inclusive . . .* by William Sutton).

[3] *Beiträge zur Theorie der Dienstunfähigkeit und Sterbensstatistik*, Berlin, 1886–91.

[4] Küttner, "Neuere Untersuchungen über die Invalidität der Steinkohlenbergleute Preussens" (*Zeitschrift für das Berg-Hütten—und Salinenwesen im Preussischen Staate*, 1888).

[5] Neison, *The Rate of fatal and non-fatal Accidents in and about Mines and on Railways*, London, 1880.

lem was relatively easy when a special group was treated, particularly if each individual could be observed. More dubious were the investigations dealing with death lists and census results, even though great care was taken. Migrations and changes of occupation often made it difficult or impossible to obtain reliable observations. The problem of the mortality of *servants* forms an example of this difficulty, a person sometimes perhaps being registered as servant and at other times, for instance, under protracted disease, being put down as "son" or "daughter." An attempt to control errors of this kind was made by Sippel with regard to a group of German servants.[1] Another attempt was made by M. Rubin in co-operation with the present author, to try to elucidate the displacements within various classes of population.[2]

Naturally investigations on subjects of a more or less complicated and disputable nature were not always un-biassed. Preconceived conceptions sometimes checked an objective treatment of the facts. This was the case with the often bitter discussion on the influence of intoxicating liquors on health, or with regard to problems concerning prostitution. Generally there was no doubt as to the correctness of the observations themselves, the controversy arising merely out of their proper interpretation. But there is in the history of statistics a fortunately isolated case of *direct falsification* of the observations. As a contribution to the discussion on vaccination against small-pox, an Austrian physician, Keller, had published (1873–6) some observations in favour of non-vaccination. These investigations seemed to have great weight, till J. Kőrösi, after the death of Keller, by a partial reconstruction of the material, proved that the observations had been falsified so as to give an entirely wrong impression.[3]

---

[1] Sippel, *Das Bamberger Dienstboten-Institut*, Bamberg, 1889.

[2] M. Rubin and H. Westergaard, *Landbefolkningens Dødelighed i Fyens Stift*, Copenhagen, 1886.

[3] J. Kőrösi, *Die Wiener impfgegnerische Schule und die Statistik, Braunschweig*, 1887. *Kritik der Vaccinations-Statistik und neue Beiträge zur Frage des Impfschutzes*, Berlin, 1889. "Neue Beiträge zur Frage des Impfschutzes," ibid., 1891.

Fortunately this record of a deliberate adulteration of statistical data stands alone.   A wrong or one-sided interpretation of statistical material is one thing, direct falsification quite another.   It cannot be denied that the former has been frequent enough even though decidedly less in the epoch here under review, when a true scientific spirit absolutely prevailed among statisticians.   But where is the guarantee that observations are not falsified?   Will it always be possible to discover frauds of this kind?   The answer to this question must be, that even if Keller's material had never been revised, even if this forgery had never been proved, his investigation would undoubtedly sooner or later have been laid aside as dubious.   In fact, in statistics, as in all science, the experience all over the world on any question will necessarily form a system, the reliability of the results being proved by their inner harmony.   Renewed investigation on the effect of vaccination would have shown that Keller's experience was not reliable; Kőrösi's own contributions in this field belonged to such a system.   But it is worth while to record a case of this nature in the history of statistics.

What has been said here of the state of mortality statistics at the close of the past century cannot claim to be exhaustive.   The literature on this subject was increasing rapidly in all countries, in official reports as in monographs.[1]   Generally it may be said that mortality statistics had now reached a point of relative completeness; the theoretic equipment was sufficient for all principal subjects, and a large material of statistical observations was at hand.   This, of course, did not mean that refinement of methods would not be welcome, so much the more because each generation will always wish to give a new form to old theories, and fresh observations would also be necessary for the confirmation or revision of previous results.   Among problems which were not yet settled was that of the influence of intoxicating liquors

---

[1] A more detailed record may be found in Westergaard, *Die Lehre von der Mortalität und Morbilität*, 2nd ed., Jena, 1901.

on health and working power; further, questions of inheritance, or of nutrition—particularly of infants, and of the results of cures for diseases, as, for instance, in tuberculosis sanatoria.

82. The theoretic insight of the statisticians not only sufficed for investigations on mortality, but with slight modifications the same methods could be applied to the whole field of vital statistics. Taking, for instance, criminality, the question may be to find the probability of a person who has been sentenced once receiving a new sentence, or in marriage statistics it may be asked what is the probability for a person of dying single, or married, or as a widower, etc. And it was now much easier to get sufficient observations. Quetelet dreamt of giving a picture of the whole human life, from the cradle to the grave, the growth of the body, the frequency of marriage, etc. But in his time it was impossible to realise all these wishes properly; only at the close of the century were the conditions more favourable. Other subjects could be added, such as income in relation to age, or unemployment, invalidity, etc. As examples may be mentioned the rate of marriage of bachelors and widowers, the correlation between the ages of bride and bridegroom, birthrates of married and unmarried women, correlated with the number of previous births, mortality in large and small families, and detailed observations on still-births. The beginning decline of the birth-rate gave fresh impulse to the discussion on population problems, and naturally deeper-going studies of birth statistics were thereby stimulated.

As has been shown above, *Sweden*, with its early development of vital statistics, ranked high with regard to statistics of *marriages and births*. As far back as 1831 it is possible in this country to calculate rates of marriage of both sexes according to age. Later the numbers were specialised so as to give the frequency of marriages of widows and widowers, and for the last decade of the past century further progress was made. As to births, the age of mothers was known since 1776 (of unmarried since 1868).[1]   An investigation concerning the Swedish Nobility was undertaken by P. Fahlbeck.[2]

Both as to marriages and births many other countries made efforts in the latter part of the century, for instance, with regard to the age-distribution of the married couples. In *Denmark*, from the year 1870 (for Copenhagen from 1878), the ages of the mothers were registered.   In the census in Copenhagen, 1880, the schedules contained details with regard to duration of marriages and the numbers of children, alive or dead.   This material was treated by M. Rubin and the present author, the result showing a remarkable difference both as to size of families in each social class and as to mortality among the children.   A supplementary investigation gave details with regard to prenuptial conceptions in the rural districts.[3]   The latter question was also taken up in *New South Wales*, where from 1893 observations were at hand as to the ages of mothers and the period elapsing from the date of marriage to the birth of the first child.   This was also the case in *Amsterdam*, and the influence of the social class was investigated in *Holland* in various places for couples entering marriage in 1897.   A. N. Kiær tried to find corresponding numbers for *Norway* (1894) by a method of sampling.[4]   In *France* the age of the parents was registered from 1892; at the census 1886 the schedules included a question concerning the number of children in the marriage, and in 1891 and 1896 in combination with the duration of the marriage.   For *Italy*, as mentioned above, the problem was taken up by E. Raseri, who tried to calculate the natality for mothers in different ages, and, further, the frequency of still-births

---

[1] Gustav Sundbärg, *Bevölkerungsstatistik Schwedens*, 1750–1900. Stockholm, 1907. Arosenius, l.c., p. 551.

[2] P. E. Fahlbeck, *Sveriges Adel.*, I–II, Lund, 1898–1902.

[3] M. Rubin and H. Westergaard, *Ægteskabsstatistik paa Grundlag af den sociale Lagdeling*, Copenhagen, 1890. In German: *Statistik der Ehen auf Grund der socialen Gliederung der Bevölkerung*, Jena, 1890.

[4] A. N. Kiær, *Statistische Beiträge zur Beleuchtung der ehelichen Fruchtbarkeit*, I–II, Christiania, 1903, p. 93; III, 1905, ibid.   This work contains a good survey of the literature on the subject.

as influenced by the age of the parents, etc.[1]   About the same time *Austria* got detailed birth statistics (from 1895), and the same was the case with *Hungary* (1897); in the latter country further contributions were made by Kőrösi, who treated observations on the existing couples (1891) and the legitimate births (1889–92), under consideration of the ages.[2]

Able contributions were made in *Germany*, for instance, in the Grand-duchy of *Oldenburg*, where valuable observations for the decade 1876–85 were published, with details concerning the age of the mothers.   R. Böckh succeeded in getting detailed material for *Berlin*, which he treated in various reports.   The ages of the husbands and wives at the census 1885, the year of marriage and the number of children were recorded.

The problem of *sex-proportion* was treated for *Saxony* by Geissler, who had a very large amount of material at his disposal.   Geissler contributed essentially to the solution of this problem.[3]   He also dealt with the question of the influence of the death of a child in a family on the chance of a subsequent conception.

It follows from what has been said that the last decades of the nineteenth century were fertile with regard to problems of marriages and births.   Several other contributions could be added, for instance, on divorces,[4] a subject which naturally began to attract attention.   In this respect Böckh's reports for Berlin were mentioned above.

[1] E. Raseri, "Les Naissances en rapport avec l'âge des parents." *Bull. de l'Institut international de statistique*, X, 1897.

[2] J. Kőrösi, "An Estimate of the Degrees of Legitimate Natality as derived from a Table of Natality compiled by the Author from his Observations made at Budapest," *Philosophical Transactions of the Royal Society of London*, Vol. 186, London, 1896.   The same author, "Zur Erweiterung der Natalitäts- und Fruchtbarkeits-Statistik," *Bull. de l'Institut international de statistique*, VI, 2, 1892.

[3] A. Geissler, "Beiträge zur Frage des Geschlechtsverhältnisses der Geborenen," *Zeitschrift des K. sächsischen Stat. Bureaus*, Jahrg. 1889, Heft 1–2; "Ueber den Einfluss der Säuglingssterblichkeit auf die eheliche Fruchtbarkeit," *ibid.*, 1885.

[4] For instance, J. Bertillon, "Le divorce et la séparation de corps," *Journ. de la Soc. de statistique de Paris*, 1884.

This period also proved rich as regards *criminal* statistics. For generations useful knowledge in this field had been available, particularly in *France* and *Belgium*. The material permitted conclusions as to the frequency in various groups of crimes, taking age, sex, etc., into consideration. But it was a serious objection, that generally the reports were not based on individual observations. It was maintained that if an individual card was filled in for each person who had been in conflict with criminal laws, many questions of great interest, particularly with regard to *recidivism*, could be solved. The problem was earnestly discussed in the nineties, for instance, at a congress in Paris, 1893. G. v. Mayr, among others, proposed a reform of this kind and his views were generally shared by statisticians.[1] As a country where judicial statistics were reformed in this direction *Denmark* may be quoted. Here observations on individual criminals were collected since 1897.[2] It was now possible to arrive at an approximate calculation of the probability of recidivism for the various groups of crime, in the form of tables of decrements corresponding to life-tables, the causes of death being replaced by crimes, and the age by the time elapsed since the first sentence. It was a drawback, however, that migrations and deaths could not easily be taken into consideration.

As to the *physical* qualities of the population, the last decades of the nineteenth century also show progress. Here, as before, military service supplied valuable material. As an instance can be quoted the measurements of recruits in *Switzerland*, the reports giving details with regard to the differences according to occupation. *Italian* military statistics also take a prominent place. It was resolved to collect very detailed observations on soldiers born 1859–63, altogether about 300,000.

---

[1] G. v. Mayr, "Zur Reform der Rückfallstatistik," *Allg. Statistisches Archiv*, III, 1894.

[2] "Strafferetsplejen i Aarene 1897–1900," *Statistisk Tabelværk*, 5 R. B. 4, Copenhagen, 1903.

Ridolfo Livi was entrusted with the work,[1] the results being published in two big volumes, 1898–1905.[2] First of all the height was measured, then the weight, width of chest, cephalic index, colour of hair and eyes, etc. Geographical distribution was taken into consideration as well as profession, and the correlation between several of the qualities was found, as, for instance, colour of hair and eyes, height and weight. Further, the anthropological data about soldiers who were taken ill or discharged on account of ill-health were considered. On the whole, the report was a rich source of knowledge, which has hardly yet been completely exhausted. In *Sweden* a corresponding though less detailed investigation with regard to conscriptions in 1897–8 was undertaken by Retzius and Fürst.[3] Several other contributions might be quoted.[4] A weighty investigation on school-children was carried out by Geissler and Uhlitzsch,[5] and many other reports appeared on the same subject. Erismann[6] treated observations on labourers in Central Russia, taking into account the influence of occupation. Thus, at the close of the nineteenth century an abundance of anthropometric observations was available. To this may be added observations with regard to heredity, for instance, Galton's investigations of the correlation of the stature of parents and their offspring.[7]

The problem of *income* and *cost of living* in various

[1] R. Livi, "Résultats obtenus du dépouillement des feuilles sanitaires des militaires des classes 1859 à 1863," *Bull. de l'Inst. international de statistique,* VII, 2, 1894.

[2] R. Livi, *Antropometria militare,* Parte I: "Dati antropologici ed etnologici," Roma, 1898. Parte II: "Dati demografici e biologici," ibid., 1905.

[3] Gustaf Retzius und Carl. M. Fürst, *Anthropologia Suecica, Beiträge zur Anthropologie der Schweden,* Stockholm, 1902.

[4] Confer, for instance, a bibliography in Hartwell, "A Preliminary Report on Anthropometry in the United States," *Bull. de l'Inst. international de statistique,* VIII, 1895.

[5] Geissler und Uhlitzsch, "Die Grössenverhältnisse der Schulkinder des Freiberger Bezirks," *Zeitschr. des k. sächsischen statistischen Bureaus,* 1888.

[6] Erismann, "Untersuchungen über die körperliche Entwickelung der Arbeiterbevölkerung in Zentralrussland," *Brauns Archiv für soz. Gesetzgebung und Statistik,* I, 1888.

[7] See, for instance, Fr. Galton, *Natural Inheritance,* London, 1889, pp. 87 *sq.*

stages of life also attracted interest.   Here, as an instance, Norwegian observations may be mentioned.   A. N. Kiær and E. Hanssen applied the method of sampling to the problem of finding the income through life of persons in various social classes.[1]   As to the *cost of living*, Le Play's school was mentioned above with its detailed investigations on individual family-budgets, and also Engel's statistics concerning *Belgian* labourers.   A large field lay open here, the difficulty mostly consisting in the condensation of the many-sided material into a system of results with that inner harmony which must be the object of every scientific investigation.

83. Going through the statistical literature of this period, we shall find that tasks which were looked upon as difficult earlier in the century could now be mastered with much greater ease.

Levasseur's and Bodio's international population statistics were mentioned above.   In a presidential address at the Statistical Society of London, Rawson W. Rawson could give an international survey of vital statistics in Europe.[2]   In many countries the statistical offices began publishing international statistics as an appendix to their year-books, or otherwise.   It was now possible to collect *more complete* and reliable international economic statistics than hitherto.   Neumann-Spallart's work in this field was ably continued by Juraschek.[3]

A prominent part in the collection of statistical knowledge was taken by the numerous statistical journals. We have seen that in several countries, such as Austria, Bavaria and Prussia, journals were published by the statistical office or in close connection with it, or a statistical association published a journal.[4]   The character of these

---

[1] A. N. Kiær and E. Hanssen, *Socialstatistik*, I–III, Bilag til den parlamentariske Arbejderkommissions Indstilling, Christiania, 1898–9.

[2] Rawson W. Rawson, "International Statistics illustrated by Vital Statistics of Europe and of some of the United States of America" (*Journ. Stat. Soc. London*, XLVIII, 1885).

[3] Juraschek, *Uebersichten der Weltwirtschaft*, Jahrgang, 1885–1902.

[4] *Quarterly Publications of the American Statistical Association*, New Series, 1888—*Journal de la Société Suisse de Statistique*, 1866—*Journal de la Société de Statistique de Paris*, 1860—

journals was varying.   Frequently they contained official communications on vital or economic statistics, sometimes also theoretic investigations, as was the case with the *Annali di Statistica*.[1]   In this connection can be mentioned various journals also of a special nature, as for instance the *Journal of the Institute of Actuaries*.   At the beginning of this century K. Pearson and others began *Biometrika* (1901) for problems of biometry.   A wider programme was offered by a journal which G. v. Mayr started in 1890.[2]

It is not without interest to follow the *Journal of the Statistical Society of London* (from 1887 the Royal Statistical Society).   As in the first period of the existence of the Society, we find many articles dealing with problems of political economy.[3]   Later, however, a division of labour took place, the *Economic Journal* having been started in 1891.   The journal of the Statistical Society was faithful to tradition as to social statistics, Charles Booth contributing various interesting investigations.[4] Old-fashioned articles also appear, as when W. A. Guy wrote on the relation between temperature and mortality.[5]   But in addition to this we meet an increasing number of investigations of a purely theoretic character. The names of F. Y. Edgeworth, Karl Pearson, G. Udny Yule and A. L. Bowley are now frequently found in the journal.   F. Y. Edgeworth, for instance, wrote on index-numbers (1883), on variations in the rates of births, deaths and marriages (1885), on the mathematical theory of banking (1888), on the statistics of examinees (in the same year), with the law of error as starting-point, on "Asymmetrical Correlation between Social Phenomena" (1894), and on the representation of statistics (1898–1900).   K. Pearson, in co-operation with Miss

---

[1] *Annali di Statistice*, Serie 2ᵃ 1878–81, 3ᵃ 1882–5, 4ᵃ 1884–1911.

[2] *Allgemeines statistisches Archiv*, 1890– .

[3] For instance, Bonamy Price (1882), "Economy and Trade."   Alfred Marshall, "Some Aspects of Competition" (1890).

[4] For instance, Charles Booth, "Condition and Occupations of the People of East London and Hackney" (1888).

[5] W. A. Guy, "On Temperature and its Relation to Mortality" (1881).

A. Lee, had an article on "Mathematical Contributions to the Theory of Evolution" (1897), but his investigations were mostly published elsewhere, as in *Philosophical Transactions*. A. L. Bowley took up the problem of wage-statistics (1895); in 1897 he wrote an article on the accuracy of an average.[1] G. Udny Yule finally wrote a paper on the history of pauperism in England and Wales (1896) and on the theory of correlation (1897), which will be mentioned below.[2]

These investigations are signs of a new tide in the evolution of statistics with fresh initiative in the field of theory.

They were met, however, by a conservative criticism from many statisticians, who wanted to follow the track which had hitherto been best known. In point of fact there was much to be done merely by trying to obtain a survey of all the statistical knowledge which was now at hand. G. v. Mayr (1841–1925) undertook the giant task of giving a critical compendium of all the results arrived at so far. While working he saw the literature swelling to an immense degree, and although he spent a long and uncommonly laborious life on his task, he did not succeed in realising the whole plan. The first two volumes appeared 1895–7, and he lived to finish the greater part of a second revised edition of these volumes.[3] The volume on economic statistics which he had planned, however, never appeared.

84. The great technical progress of statistics during the last decades of the nineteenth century had made it much easier than before to collect and to master a vast material and to avoid inaccuracy in the observations. This will help to explain the general views of G. v. Mayr

---

[1] A. L. Bowley, "Relations between the Accuracy of an Average and that of its Constituent Parts" (1897).

[2] G. Udny Yule, "Notes on the History of Pauperism in England and Wales from 1850" (1896); "On the Theory of Correlation" (1897); "An Investigation into the Causes of Changes in Pauperism in England" (1899).

[3] G. v. Mayr, *Statistik und Gesellschaftslehre.* I—"Theoretische Statistik," Freiburg, 1895; II—"Bevölkerungsstatistik," ibid., 1897; III—"Moral-statistik," Tübingen, 1909–17.

and his school, which also in the present century are shared by many statisticians.   It was their aim to have at their disposal as many *direct* observations as possible, covering the whole ground which they had to explore. They aimed at getting, by extensive use of census-taking and other enumerations, as complete a picture as could be obtained.   This, as a rule, only called for elementary methods and they felt no need for complicated mathematical formulæ.   Only in special cases would it be advisable to consider the oscillations around the mean; as a rule the large volume of the material would be a sufficient guarantee of the reliability of the conclusions. Concrete problems might invite the use of applied mathematics, but this would chiefly belong to the field of "political arithmetic." [1]

According to the general view of the school it should be the task of the statisticians to give a thorough description of the state of things in their inner causal connection, and aiming as they did at the widest possible extension of direct enumeration, they would generally feel disinclined to try *indirect* methods.[2]   Undoubtedly all statisticians will agree in preferring direct observations, if available, to indirect conclusions, but it was maintained by various statisticians that in many cases it was necessary to use indirect methods.   *Sampling*, therefore, presented itself as a natural way out, where it proved impossible to provide observations from the whole field.   A thorough theoretical foundation for this method had to be postponed to the present century, but in the period here concerned we meet various energetic attempts at an application of it, and on the whole tolerably reliable results were obtained, even though these attempts did not escape vivid protests from other statisticians.   A. N. Kiær has already been mentioned in this connection.

As an example the above-mentioned report upon in-

---

[1] *Statistik und Gesellschaftslehre*, I, pp. 20, 118–19.

[2] G. v. Mayr, *Deutsche Arbeiterstatistik*, pp. 133–4; *Allg. Statistisches Archiv*, 3, 1894.

come and wealth in *Norway* in 1891 can be quoted.[1] Some of the towns and a number of parishes in the rural districts, as far as possible representative of the whole population, being selected, and all persons of certain ages (17, 22, 27, etc.) and with certain initial letters in their names, were recorded, in five social classes and according to marital condition. Another application of the method was used by A. N. Kiær and E. Hanssen for a parliamentary labour commission, with the object of finding the correlation between age, income, invalidity and occupation, etc., by means of sampling.[2] In the capital a number of streets was chosen, as fairly as possible representing the various classes of population; in the provincial towns and in the rural districts a corresponding selection was made. In a number of houses chosen at random—for instance, every tenth house in populous streets in the capital—the adult persons present in 1894 were recorded, and on the basis of the results conclusions were drawn as to the total population. Even though these keen experiments invited criticism, it could not be denied that the underlying principle was as sound as the attempt of Laplace at finding the population of France at the beginning of the century, not to speak of anthropometric observations or the numerous monographs on the influence of occupation on health. Naturally the method of sampling found application in *agricultural* statistics. It was used in *Denmark* with regard to the harvest of 1901,[3] and a few years after P. Mayet applied it to the live stock in *Baden*.[4]

[1] "Indtægts- og Formueforhold i Norge," *Norges officielle Statistik*, III, No. 255, Kristiania, 1897. A. N. Kiær, "Die repräsentative Untersuchungsmethode," *Allg. Statistisches Archiv*, 5, 1899–1900. The same author, "Observations et expériences concernant des dénombrements représentatifs." *Bull. de l'Institut international de statistique*, IX, 1895.

[2] *Socialstatistik*, Bilag til den parlamentariske Arbejderkommissions Indstilling, Kristiania, 1898–9.

[3] "Høsten i Danmark i Aaret, 1902," *Statistiske Meddelelser*, 4 R., XIII, 1903.

[4] P. Mayet, "Stichproben—Erhebungen in der Zwischenzeit zwischen grossen Vollzählungen längerer Periodizität," *Bull. de l'Institut international de statistique*, XIV, 1904.

The method of sampling may be considered as a short-cut, where otherwise it will be difficult to obtain results. Other methods—intended as labour saving—were also subjected to opposition. The *standard calculations* have been mentioned above. By this method a brief expression should be obtained of the influence of certain causes, after elimination of disturbing factors, for instance the different age-distribution. Although evidently summary calculations of this kind could not dispense with a detailed study of the observations, many statisticians found them useful for a preliminary investigation and sometimes there may have been a temptation to go somewhat further, ascribing to the method more value than it actually possessed. This will explain the discussion on this problem at the Sessions of the International Statistical Institute. In the session 1891 in Vienna, W. Ogle (who had applied standard calculations to the statistics of the influence of occupation on health) proposed to establish an international use of a standard population with fixed sex- and age-distribution, in the calculation of marriage, birth- and death-rates. A committee was appointed to deal with this proposal, as also with a corresponding one by Kőrösi. The problem was discussed at the session in Berne, 1895, and for standard calculations of this kind, the use of the population of *Sweden*, with its long history of vital statistics, was recommended. But at the session 1897 in St. Petersburg criticism arose, and L. v. Bortkiewicz expressed most energetically the discord existing among the statisticians in regard to this problem.[1] This problem was therefore also left unsettled for debates in the present century.

It was often acknowledged, however, that index-calculations had their mission. This was the case with *family-budgets*. Here Engel proposed to take the sex and ages of the members of the household into consideration by means of a scale, allotting to each member of

[1] *Bull. de l'Institut international de statistique*, XI, 1, 1899, pp. 174 *sq.*

the family a weight, increasing with age to 3·5 for males and 3·0 for females, and expressing the whole relative power of consumption for the household in units, which, in honour of Quetelet, he would call Quets.[1] On the whole the *index-numbers* in the statistics of prices were generally more and more accepted in this period, even though theory added little to what had already appeared. As one of the contributions to the solution of the problem an article by Edgeworth[2] (1883) can be quoted, in which he prefers the arithmetic calculation to the method which Jevons recommended.[3]

In this connection another problem deserves mention which has been much discussed in the present century, but which in the period concerned was only treated occasionally, viz. that of the *economic periods*. As a natural solution it was recommended—though not with general assent—to use a trigonometric formula in order to express the economic movements.[4]

Pareto's formula for the distribution of incomes in this period attracted great interest.[5] This is not the place to discuss the value of this theory from the point of view of political economy, but it had a just claim to be accepted as a useful formula of interpolation for deliberations on the scale of income taxes, statistics of income being generally not so specialised that

---

[1] E. Engel, *Der Kostenwert des Menschen*, Berlin, 1883. " Die Lebenskosten belgischer Arbeiter-Familien früher und jetzt," *Bull. de l'Institut international de statistique*, IX, 1895, 1, pp. 4–8.

[2] Edgeworth, "On the Method of ascertaining a Change in the Value of Gold" (*Journ. Stat. Soc. London*, XLVI, 1883).

[3] As to theoretic advantages of this method, see Westergaard, *Theorie der Statistik*, Jena, 1890, pp. 219 *sq.*

[4] J. H. Poynting, "A Comparison of the Fluctuations in the Price of Wheat and in the Cotton and Silk Imports into Great Britain" (*Journ. Stat. Soc.*, XLVII, 1884).

[5] Pareto, *La courbe de la répartition de la richesse*, Lausanne, 1896. Edgeworth, "Supplementary Notes on Statistics" (*Journ. Stat. Soc.*, LIX, 1896); confer J. C. Stamp, "A new Illustration of Pareto's Law" (*Journ. Stat. Soc.*, LXXVII, 1914).

If y is the number of persons with incomes of x or more, while a and b are constants, the formula is: $x^a y = b$.

problems of this kind could be solved by direct observation.

From this point of view the case is analogous to the
problem arising where a certain class of society is distributed into a few classes of age, and a more detailed
division is required.[1]

85. The problem of the *accuracy* of the observations
often attracted attention of course, as, for instance, errors
as to age at a census or through the registration of
deaths.   Various investigations from the last decades of
the century can be recorded.   For example, those by
Würzburger with regard to the census in Saxony,[2] and
Oehrn concerning the ages of death in *Latvia*.[3]   The
Norwegian Central Bureau of Statistics made inquiries
as to the ages of persons above 95, in order to correct
the census results from 1891.[4]   Even though technical
progress removed various sources of error, for instance,
by registration of age as well as of birth-year at the
census, some slight irregularities were evidently left,
and it was natural to try to adjust the observations by
some mechanical process or by means of a mathematical
formula, for instance, in order to get a smooth series
of rates of mortality according to age.   As we have
seen, this problem of *adjustment* had found a scientific
solution already at an earlier stage, chiefly through
application of the method of least squares.   Naturally
several investigations on this problem appeared in the
last decades of the past century.   It was proved thereby
that practically useful results could be obtained by relatively elementary methods; for instance, by means of a
graphic or mechanical adjustment through application
of various corrections, by the method of least squares,

---

[1] Westergaard, "Die Gliederung der Bevölkerung nach Gesellschaftsklassen,"
Allg. Statistisches Archiv, 4, 1895–6.

[2] E. Würzburger, "Zur Frage der Genauigkeit der Volkszählungen," *Jahrb.
für Nationalök und Stat.*, 1896, I.

[3] E. Oehrn, *Biostatik dreier Landkirchspiele Livlands in den Jahren 1834–81*,
Dorpat, 1883.

[4] Westergaard, "Mortality in Extreme Old Age," *The Economic Journal*,
IX, 1899.

or otherwise. By these adjustments Makeham's formula frequently rendered good service.[1]

It is not within the plan of the present treatise to describe the progress of the *theory of probabilities* in the period concerned, unless a direct influence upon the evolution of statistics can be traced. We have seen that the distribution of observations around the mean had naturally attracted statisticians, and here "the normal law of error," the distribution according to the exponential or rather the binomial formula, had a prominent part. Fr. Galton invented an apparatus which in a mechanical way formed a binomial symmetrical distribution[2]; it was later on generalised by K. Pearson so that it could be applied to any value of the probability.[3] Results of various games gave good contributions.[4] Many anthropometric observations more or less show an approximation to the binomial law, Lexis having proved that observations on sex-proportion were in good harmony with it. Less obvious was the harmony in other fields of vital and social statistics, not to speak of economic statistics. But it could be shown that an approximation to the law could frequently be obtained rather easily by a division of the observations.[5] An interesting contribution was made by L. v. Bortkiewicz.[6]

[1] See, for instance, Th. B. Sprague, "The Graphical Method of adjusting Mortality Tables" (*Journ. Inst. Act.*, XXVI, 1897). J. Karup, "Ueber eine neue mechanische Ausgleichungsmethode," *Transactions of the second International Actuarial Congress*, 1899. E. Blaschke, *Die Methode der Ausgleichung von Massenbeobachtungen*, 1893. W. Lazarus, *Zur deutschen Lebensversicherungs-Sterblichkeitstafel Ass. Jahrb.*, VI, Wien, 1885. G. King, *Institute of Actuaries' Textbook*, II, London, 1887, pp. 78 *sq.* As to G. F. Hardy's adjustment, see *The British Offices' Life-Tables*, 1893; *An Account of the Principles and Methods*, London, 1903.

[2] Fr. Galton, *Natural Inheritance*, London, 1889, p. 63.

[3] K. Pearson, "Skew Variation in Homogeneous Material, I," *Phil. Trans. Royal Soc.*, Vol. 186, 1895, § 1.

[4] See, for instance, K. Pearson, *The Chances of Death*, I, London, 1897, p. 13. Westergaard, *Grundzüge der Theorie der Statistik*, Jena, 1890, pp. 21 *sq.*

[5] Westergaard, l.c., pp. 43 *sq.*

[6] L. v. Bortkiewicz, *Das Gesetz der kleinen Zahlen*, 1898. Independently A. L. Bowley arrived at the same result, *Elements of Statistics*, 4th ed., 1920, pp. 284 *sq.*

He found that in several cases, where the probability of an event was very small, the numbers of observations were surprisingly regular. By a mathematical analysis of the theoretic distribution he gave a natural explanation of this experience, the numbers being in good harmony with theory, so that to speak of the *permanence of small numbers* was justified.

Generally the opinion of Edgeworth was shared by his contemporaries, that "where possible we should employ the normal law of frequency."[1] In order to realise this it would often be required to prepare the material carefully, dissecting the observations in such a way that the effect of the various causes could be seen. Any deviation from the normal law of frequency would be an indication that there were separate causes in force, and it was consequently the duty of the statistician to explore these causes. This, then, gave the modern statistician a clear programme for his work. As a matter of course this dissection would frequently be extremely difficult or even impossible, even where it was evident that some strong influences were prevailing. It might, for instance, be an effect of these influences that the observed curve of frequency was double-humped with *two* maxima (whereas the normal curve has only one), as the case might be with the height in a population constituted of two separate types. K. Pearson took up the problem of a dissection of frequency-curves in order to disclose separate types. The result may be of interest even though the experiment must always have a hypothetical character. As an instance he treated some observations on the cephalic index of *South-Germans* in primitive times and at present. O. Ammon had interpreted these numbers as showing a selection which had gradually reduced the dolicocephalic elements. By breaking up the material into two components Pearson arrived at the result, that the old inhabitants were a mixed population, one

---

[1] F. Y. Edgeworth, *The Representation of Statistics by Mathematical Formulæ,* 1900, p. xix.

part having exactly the same character as the modern Germans.[1]

Whether a dissection of this kind was possible or not it was an important problem to study the form of the frequency-curves, asymmetric as well as symmetric, including the above-named double-humped curves. Edgeworth, for instance, took the normal distribution as starting-point, proposing a method of "translation" from a hypothetical group obeying the normal law,[2] or attempts were made at combining two halves of different normal curves.[3] The *Danish* mathematician T. N. Thiele used a system of values from the observations to characterise the frequency-curves, symmetric as well as asymmetric.[4] K. Pearson would solve the problem by means of a general equation, which by change of constants would lead to various forms of the frequency-curve, including the normal curve as a special case.[5]

86. There is another important subject which gave statisticians fresh impulses at the close of the past century, viz. the *theory of correlation* which since then has had a prominent part in the statistical literature, particularly in *England and America*, and most specially in biometry. The elements of the theory are contained in the above-quoted investigation from 1846 by the French astronomer Bravais (art. 54), but in its present form it is due to English statisticians who developed the theory without at first noticing Bravais' contribution. During his studies on heredity Fr. Galton, 1877, measured the size of sweet-peas, and found a peculiar

---

[1] K. Pearson, "Contributions to the Mathematical Theory of Evolution," *Philos. Trans.*, Vol. 185, 1894, p. 107. O. Ammon, *Die natürliche Auslese beim Menschen*, Jena, 1893, pp. 66 *sq.*

[2] Edgeworth, l.c., pp. 6 *sq.*

[3] See, for instance, the posthumous work by G. Th. Fechner, *Kollektiv-masslehre*, edited by Fr. Lipps, Leipzig, 1897.

[4] T. N. Thiele, *A mindelig Iagttagelseslære*, Copenhagen, 1889. *Elementær Iagttagelseslære*, ibid., 1897. *Theory of Observations*, London, 1903.

[5] K. Pearson, "Contributions to the Mathematical Theory of Evolution. Skew Variation in Homogeneous Material," *Phil. Trans.*, Vol. 186, 1895. His theory is easiest accessible in W. Palin Elderton, *Frequency Curves and Correlation*, London, 1906.

"*regression*" of daughter seed compared to mother seed towards the general mean.[1] Eight years later he returned to the problem. In connection with the health exhibition of 1884 he initiated an anthropometric laboratory [2] where he collected observations on the inheritance of stature in man and in 1885–6 and later he published various articles on the results. Comparing, for instance, the stature of fathers and sons, he found an evident relation, tall fathers having on an average tall sons, etc., but there was in the latter group of observations a step backwards, a "regression" to the mean. These results could be represented by a straight line (a line of regression) showing how much the average stature of the adult offspring was reduced compared to the parental stature. In dealing with this problem J. D. Hamilton Dickson reached some of Bravais' results. Also Edgeworth discussed the subject.[3] Later Pearson and Yule took up the question. Pearson gave the theory its present form with its *coefficient of correlation* and its coefficients of regression,[4] and Yule treated the subject theoretically as well as practically, taking particularly the pauperism in England as starting-point.[5]

By these investigations the conception of the significance of the line of regression changed. In fact, the same correlation between two groups of observations as

---

[1] Fr. Galton, *Memories of my Life*, 1908, pp. 300 *sq.* K. Pearson, *The Life, Letters and Labour of Francis Galton*, 1914–30, III A, 1930, pp. 6 *sq.*

[2] Fr. Galton, "Some Results of the Anthropological Laboratory," *Journ. of the Anthropological Institute*, XIV, 1885; "Regression towards Mediocrity in Hereditary Stature," ibid., XV, 1886; "Family Likeness in Stature," *Proceedings Royal Society*, XL, 1886 (with an Appendix by J. D. Hamilton Dickson); "Co-relations and their Measurement," ibid., XLV, 1889; Natural Inheritance, 1889.

[3] F. Y. Edgeworth, "On Correlated Averages," *Phil. Mag.*, 5th Ser., XXXIV, 1892.

[4] K. Pearson, "Mathematical Contributions to the Theory of Evolution, Regression, Heredity and Panmixia," *Phil. Trans.*, Vol. 187, 1897.

[5] G. Udny Yule, "On the Theory of Correlation," *Journ. Stat. Soc.*, LX, 1897; "On the Correlation of total Pauperism with Proportion of Out-Relief," *The Economic Journal*, V, 1895; VI, 1896; "An Investigation into the Causes of Changes in Pauperism in England chiefly during the last two Intercensal Decades," *Journ. Stat. Soc.*, LXII, 1899.

with regard to heredity could be found in many other fields, as for instance between old-age pauperism and proportion of out-relief, age at marriage of brides and bridegrooms, stature and weight, etc. This again would influence the interpretation of Galton's observations on regression from father to son, the question being whether the regression partly, or wholly, was a consequence of the unavoidable mixture of types, or whether a real stepping backwards had taken place.[1]

Strictly speaking, the theory of correlation did not introduce any new principle, it being entirely based on well-known theorems of the Calculus of Probabilities. But it gave statisticians easily-handled formulæ, containing, so to speak, a complete programme for their investigations, and the vast application of these formulæ to biometrical and other problems gave the statistical literature to a large extent a new appearance. On the other hand, it cannot be denied that these formulæ contained a danger, tempting as they might to a too mechanical treatment of the observations, and so to speak increasing the distance .between the observer and the facts, whereas the old-fashioned methods had the advantage of keeping the observations themselves more in view, and therefore often made it easier to draw safe conclusions from the material. It was left to the coming decades to find the balance between old and new methods. It may be added that naturally the theory of correlation with its numerous theoretical problems attracted mathematicians, who were stimulated to deeper investigations, the latter again proving to be a profit to statistics. Here again, at the close of the past century, we meet rich possibilities of further evolution.

Summing up what has been said in the preceding chapters, it may well be maintained that on the threshold of the new century the horoscope for the evolution of statistics was a favourable one. As we have seen, several important subjects in vital and economic statistics had

---

[1] Westergaard und H. C. Nybølle, *Grundzüge der Theorie der Statistik*, 2nd ed., Jena, 1928, pp. 279 *sq.*

come more to the front, for instance as to births, wages and prices, and the scientific equipment was fully sufficient to cope with this extension of problems. It was much easier than before to collect observations, and the technical methods were in rapid development, theory being at the same time in a constant evolution. And a critical sense was awake which allowed no statistical field to be untouched till no doubt as to the results was left. All this, in fact, has characterised the statistical literature in the present century.

INDEX OF NAMES

Bumsted, 214
Burch, J. van der, 64

Caetano de Lima, L., 12–3
Caird, J., 198
Cantor, M., 15, 100
Cardan (Cardanus), H., 100
Carleson, E., 57
Caspar, W., 39
Casper, J. L., 129, 154–6
Chahrol, G. J. G., 115
Chalmers, G., 39
Châteauneuf, Benoiston de, 155–6
Cheysson, E., 247
Child, J., 39
Civiale, J., 148–9
Cless, G. P., 158
Colbert, J. B., 49
Conring, H., 6–7, 9, 11, 14
Cotes, R., 111
Cournot, A. A., 149
Coxe, Tench, 121
Craigie, P. G., 192, 198, 247
Cramer, G., 107, 109
Crome, A. F. W., 13
Cummings, J., 86, 120, 238, 241

Dael, J. van, 25–7
Darwin, Charles, 226
Davenant, Charles, 38–42, 44, 77
Davies, Griffith, 132, 134
Day, A., 213
Demonferrand, F., 160–1
Denis, H., 247
Deparcieux, A., 58, 61–3, 71, 74, 76
Deparcieux, jun., 62
Derham, W., 70–1
Deslandes, 32
Deuchar, D., 27
Diannyère, A., 97–99
Dieterici, K. F. W., 217
Double, F. J., 148
Douglas, G., 252
Drobisch, M. W., 233–4
Ducpetiaux, E., 161
Dupré de Saint-Maur, 68

Duvillard, E. E., 94–5, 115, 127
Duyn, N., 69

Easton, J., 90
Eden, Sir F. M., 84–5
Edge, G., 119, 144
Edgeworth, F. Y., 230–1, 261, 266, 269–71
Edmonds, T. R., 131
Elderton, W. Palin, 270
Elvius, P., 54–5, 57, 61
Elzevir, A. and B., 6
Emminghaus, A., 216
Enestrøm, G., 27
Engel, E., 117, 179–80, 182, 217, 240, 246–7, 260, 265–6
Erismann, F., 259
Escherich, 211
Espine, Marc d', 152–3
Eulenburg, 179
Euler, L., 68, 73, 88, 108, 126, 160, 218
Expilly, Abbé d', 79

Fahlbeck, P. E., 256
Falkenburg, P., 237
Farr, W., 137, 144, 147, 161, 183, 209, 213–6, 219, 222, 250–1
Farren, E. J., 163
Faure, F., 48–50, 52, 114, 186, 242
Fechner, G. T., 270
Fénélon, F., 52, 98
Fenger, E., 161
Fermat, P., 102
Ferri, E., 233–4
Ficker, A., 178
Finlaison, A. G., 162, 216
Finlaison, J., 134–5, 216, 220
Forbonnais, V. de, 49, 51
Fourier, J. B. J., 114–6
Franck, S., 6
Frederick the Great, 116
Fuchs, C. H., 157–8
Fürst, C. F., 259

Gabaglio, A., v, 234
Galileo, G., 100–1